PENGUIN BOOKS

FARM CITY PS - A SD - O21

Novella Carpenter grew up in rural Idaho and Washington State. She studied biology and English at the University of Washington, where she had many odd jobs, including assassin bug handler and 16-millimeter film projectionist. After moving to California, she attended UC Berkeley's Graduate School of Journalism where she studied with Michael Pollan. Her writing has appeared on Salon.com and sfgate.com and in *Mother Jones* and *Food and Wine*. Her adventures in urban agriculture began with honeybees and a few chickens, then some turkeys, until she created an urban homestead called GhostTown Farm near downtown Oakland, where she and her boyfriend, Bill, live today. Her blog documents the latest happenings on the farm: www.novellacarpenter.com.

Praise for *Farm City* by Novella Carpenter

"*Farm City: The Education of an Urban Farmer* is easily the funniest, weirdest, most perversely provocative gardening book I've ever read. I couldn't put it down. . . . Her tone is clear, relaxed and amicable; she is hilarious in describing the foibles of her friends and her '70s-era hippie parents. . . . As she contemplates the place of her garden in the greater scheme of life, she shows what she's capable of and the writing soars." —*The New York Times Book Review*

"I had a feeling I might like this memoir when I came upon its first sentence, a gentle twist on the opening of Isak Dinesen's *Out of Africa*. Here is Novella Carpenter: 'I have a farm on a dead-end street in the ghetto.' But I didn't truly fall in love with *Farm City* until I hit page 38. That's when the bees that Ms. Carpenter has purchased from a mail order company arrive at her post office in Oakland, California. A panicked postal employee calls, begging her to pick them up because they're attracting other bees and 'freaking everyone out.' . . . Fresh, fearless and jagged around the edges, Ms. Carpenter's book, an account of how she raised not only fruit and vegetables but also livestock on a small, scrubby abandoned lot in Oakland, puts me in mind of Julie Powell's *Julie & Julia* and Elizabeth Gilbert's *Eat, Pray, Love*. . . . Like Ms. Powell and Ms. Gilbert, Ms. Carpenter is very, very funny. . . . *Farm City* is filled with terrific stories. But as it strides artfully along, you begin to see that Ms. Carpenter has other things, even a larger argument, on her mind." —*The New York Times*

"If you think the local food movement is getting a tad precious, then you'll relish *Farm City*. Novella Carpenter's captivating account of the funky little farm she created on an abandoned lot in a rough section of Oakland puts a whole new twist on the agrarian tradition in America: she's going for a mind-meld of Fifty Cent and Wendell Berry, or an inner-city version of the *The Egg and I*—if you can conceive of such a thing without your head exploding. By turns edgy, moving and hilarious, *Farm City* marks the debut of a striking new voice in American writing."
    —Michael Pollan, author of *The Omnivore's Dilemma* and *In Defense of Food*

"*Farm City* is Carpenter's delectable story of how she turned a 'ghetto squat lot' in Oakland, California, into a working urban farm."
    —*O, The Oprah Magazines* (25 Books You Can't Put Down)

"You'll laugh along the way . . . especially during the July she spends eating only what she grows, confessing: 'I would have killed for a bag of red-hot Cheetos.'"
    —*People*

"I love this book in a way that makes adjectives seem puny and inadequate. It is just flat-out amazing (see? lame)—all that [Carpenter] did, the place that she did it in, the things she grew and raised and loved and killed and ate and learned. And how deeply and utterly it paid off—both for her and for anyone fortunate enough to be reading this book. [Carpenter] made me want to get up off my ass and go buy a pig! Plant heirloom watermelons! Build a chick brooder! Fortunately for all involved, the book was too good to put down and I just kept on reading instead."
    —Mary Roach, author of *Stiff* and *Bonk*

"Carpenter's adventurous memoir . . . offers a contemporary restaging of the agrarian American dream. . . . In doing so, *Farm City* offers a refreshing take on the sustainable food movement—introducing the ethical and logistical ambiguities involved in food choices without too much of the ethical high ground cultivated in many back-to-the-land primers, diet guides or muckraking exposés. . . . The fast-paced account of the day-to-day drama unfolding in one backyard in Oakland makes *Farm City* more than just a whimsical, next-generation hippie farm in the ghetto and transforms Carpenter's personal experience into a broader, more engaging inquiry into our culture's complex relationship with food."   —*San Francisco Chronicle*

"I love this book, and I love Novella Carpenter. She is insane in the best kind of way and has taught herself how to write as well as she has how to farm in a ghetto. I finished the book grateful to this uncommon individual for the choices she's made. *Farm City* is an astonishing story, effortlessly told—a wonderful debut by a writer I hope to hear a lot more from in the future."
    —Michael Ruhlman, author of *The Soul of a Chef* and *The Elements of Cooking*

"In this utterly enchanting book, food writer Carpenter chronicles with grace and generosity her experiences as an 'urban farmer.' . . . Consistently drawing on her Idaho ranch roots and determined even in the face of bodily danger, her ambitions led to ownership and care of a brace of pigs straight out of E. B. White. She chronicles the animals' slaughter with grace and sensitivity, their cooking and consumption with a gastronome's passion, and elegantly folds in riches like urban farming history. Her way with narrative and details, like the oddly poetic names of chicken and watermelon breeds, gives her memoir an Annie Dillard lyricism, but it's the juxtaposition of the farming life with inner-city grit that elevates it to the realm of the magical."   —*Publishers Weekly* (starred review)

"If you think that urban farming requires an estate-sized abode, think again. [Carpenter's] story of farming complete with livestock amidst the mean streets of Oakland is hysterical and uplifting. By the time you're finished reading, you'll be sowing seeds in the cracks of concrete around your house or raising goats in your driveway."   —Evan Kleiman, host of *Good Food* on KCRW and kcrw.com

"Carpenter's bonding with her plants and animals evokes many an emotional moment, but she never succumbs to maudlin sentimentality. Her transparent prose mirrors her urban farm: chockablock, nontraditional, and unkempt, yet intelligently real and utterly engrossing."   —*Booklist*

"Novella Carpenter's passion for local food, her resourceful attitude, and her wry yet humble sense of humor provide not just an informative manual for the urban homesteader, but also a refreshing and highly entertaining read. And while the strategies outlined in *Farm City* might not be for everyone, they certainly provide a working example of how much can be done in one urban neighborhood. Just imagine a world where each of us tried this hard to provide our own food."
          —Heather C. Flores, author of *Food Not Lawns: How to Turn Your Yard into a Garden and Your Neighborhood into a Community*

FARM CITY

THE

EDUCATION

OF AN

URBAN FARMER

NOVELLA CARPENTER

PENGUIN BOOKS

THE PENGUIN PRESS
Published by the Penguin Group
Penguin Group (USA) Inc., 375 Hudson Street, New York, New York 10014, U.S.A.
Penguin Group (Canada), 90 Eglinton Avenue East, Suite 700, Toronto,
Ontario, Canada M4P 2Y3 (a division of Pearson Penguin Canada Inc.)
Penguin Books Ltd, 80 Strand, London WC2R 0RL, England
Penguin Ireland, 25 St Stephen's Green, Dublin 2, Ireland (a division of Penguin Books Ltd)
Penguin Group (Australia), 250 Camberwell Road, Camberwell,
Victoria 3124, Australia (a division of Pearson Australia Group Pty Ltd)
Penguin Books India Pvt Ltd, 11 Community Centre, Panchsheel Park, New Delhi – 110 017, India
Penguin Group (NZ), 67 Apollo Drive, Rosedale, North Shore 0632,
New Zealand (a division of Pearson New Zealand Ltd)
Penguin Books (South Africa) (Pty) Ltd, 24 Sturdee Avenue,
Rosebank, Johannesburg 2196, South Africa

Penguin Books Ltd, Registered Offices:
80 Strand, London WC2R 0RL, England

First published in the United States of America by The Penguin Press,
a member of Penguin Group (USA) Inc. 2009
Published in Penguin Books 2010

1   3   5   7   9   10   8   6   4   2

Grateful acknowledgment is made for permission to reprint "The Arrival
of the Bee Box" from *Ariel* by Sylvia Plath. Copyright © 1963 by Ted Hughes.
Reprinted by permission of HarperCollins Publishers.

THE LIBRARY OF CONGRESS HAS CATALOGED THE HARDCOVER EDITION AS FOLLOWS:
Carpenter, Novella, 1972–
Farm city : the education of an urban farmer / Novella Carpenter.
p. cm.
ISBN 978-1-59420-221-6 (hc.)
ISBN 978-0-14-311728-5 (pbk.)
1. Urban agriculture.   I. Title.
S494.5.U72C37 2009
630.9173'2—dc22   2008054666

Printed in the United States of America
Designed by Stephanie Huntwork

To my mother,

Patricia Ann Carpenter—you are my inspiration

PART I

TURKEY

CHAPTER ONE

❋

I have a farm on a dead-end street in the ghetto.

My back stairs are dotted with chicken turds. Bales of straw come undone in the parking area next to my apartment. I harvest lettuce in an abandoned lot. I awake in the mornings to the sounds of farm animals mingled with my neighbor's blaring car alarm.

I didn't always call this place a farm. That didn't happen until the spring of 2005, when a very special package was delivered to my apartment and changed everything. I remember standing on my deck, waiting for it. While scanning the horizon for the postal jeep, I checked the health of my bee colony. Honeybees buzzed in and out of the hive, their hind legs loaded down with yellow pollen. I caught a whiff of their honey-making on the breeze, mixed with the exhaust from the nearby freeway. I could see the highway, heavy with traffic, from the deck.

I noticed that three bees had fallen into a watering can. As their wings sent out desperate ripples along the water, I broke off a twig from a potted star jasmine and offered it to the drowning insects. One bee clambered onto the stick and clung to it as I transported her to the top of the hive. The next bee did the same—she held fast to the twig like a passenger gone overboard, clutching a lifesaver. Safe atop the hive, the two soggy bees opened their wings to the morning sunlight. Once dry and warm, they would be able to fly again.

Just to see what would happen, I lifted the final rescuee to the entrance of the hive instead of the top. A guard bee stomped out from the dark recesses of the brood box. There's always one on vigil for disturbances, armed and ready to sting. As the guard bee got closer to the wet one I braced myself for a brutal natural history lesson.

The waterlogged bee started to right herself as she waved a soggy antenna. Another guard bee joined the first, and together they probed the wet bee. She couldn't have smelled of their hive anymore, which is how most bees recognize one another. Nonetheless, the guards began to lick her dry.

"Hey! Hey!" a voice yelled.

I peered down to the end of our dead-end street.

A new car, a silver Toyota Corolla, had arrived on 28th Street the night before, probably the victim of a joyride—Corollas are notoriously easy to start without a key. Local teenagers steal them and drive around until they run out of gas. Already the car had lost one wheel. By nightfall, I predicted, it would be stripped completely.

Amid the jumble of abandoned cars and trash and the shiny Toyota Corolla, I made out the figure of the man who was yelling. He waved vigorously. Bobby.

"Morning, sir!" I called and saluted him. He saluted back.

Bobby lived in an immobilized car. He switched on his television, which was mounted on top of one of the other abandoned cars. An orange extension cord snaked from a teal-colored house at the end of the block. The perky noise of Regis and Kathie Lee joined the sound of the nearby traffic and the clattering trundle of the San Francisco Bay Area's subway, BART, which runs aboveground next to the highway.

Just then, a monk came out of the Buddhist monastery across the street from my house and brought Bobby a snack. The monks will feed anyone who is hungry. Next to the fountain in their courtyard there's a giant alabaster statue of a placid-faced lady riding a dragon: Kuan Yin, the goddess of compassion. My bees loved to drink from the lotus-flower-filled fountain. I often watched their golden bodies zoom across 28th Street, at the same height as the power lines, then swoop down behind the temple's red iron gates.

The monk who handed Bobby a container of rice and vegetables was female, dressed in pale purple robes, her head shaved. Bobby took the food and shoved it into a microwave plugged in next to the television set. Nuked his breakfast.

I heard the clattering sound of a shopping cart. A can scrounger. Wearing a giant Chinese wicker hat and rubber gloves and carrying a pair of tongs, she opened our recycling bin and started fishing around for cans. She muttered to herself in Chinese, "Ay-ya."

I watched as Bobby jogged over to her. I had never seen him run before. "Get out of here," he growled. His territory. She shook her head as if to say she didn't understand and continued fishing. Bobby butted her with his belly. "I said *get*," he yelled. She scurried away, pulling her cart after her. Bobby watched her retreat.

Then, when she was almost around the corner, as if he felt bad, Bobby put his hands to his mouth and yelled, "I'll see you at the recycling center!" Just a few blocks away, the center paid cash by the pound for metal. Chuckling to himself, Bobby glanced up at me on the deck and flashed me a mostly toothless smile.

This place, this ghetto of Oakland, California, brings out the best and the worst in us.

Bored of waiting around outside, I headed back inside my apartment. A fly strip dangled from the ceiling, and ripped feed bags piled up near the door. A black velour couch my boyfriend and I found in the street sagged in the corner.

I guess the neighborhood brings out the best and worst in me, too. Sure, my chickens lay eggs—but the flock has spawned an occasional rooster that crowed loudly and often, starting at 4 a.m. Bees do result in honey and wax and better pollination—but they have also stung people from time to time. The garden: verdant cornucopia on one hand, rodent-attracting breeding ground on the other.

I flopped onto the couch and read the chalkboard tally that hung near the door:

4 chickens
30,000 bees [approximately]
59 flies
2 monkeys [me and my boyfriend, Bill]

That tally was about to change.

A long-debunked scientific theory states that "ontogeny recapitulates phylogeny." Basically that means that the order of development in an embryo indicates its evolutionary development—for example, a human embryo first looks like a fish because we evolved from fish. When Bill and I first moved from Seattle to Oakland, I was reminded of that theory, because somehow we ended up re-creating our old life in the exact same order as we had created it in Seattle. The first year in Oakland, we built the garden; the second year, we got the honeybees and then the chickens. In this, our third year of development, it was time to evolve to the next level.

Out of the corner of my eye, I watched through the window as the postal jeep turned down our street and pulled to a stop in front of our house. A man dressed in wool shorts hopped out, holding an air-hole-riddled box in his arms. I bounded downstairs. My neighbor Mr. Nguyen, who lived one floor below me, was sitting outside on the porch, smoke and steam from his morning cigarette and Vietnamese coffee wafting up together in the crisp spring air. In his sixties, Mr. Nguyen dyed his graying hair black, wore button-down dress shirts, and was surprisingly sprightly. He set down his coffee, stubbed out his cigarette, and walked into the street with me to receive the package.

The postal worker made me sign an official-looking piece of paper before he would hand me the box. It peeped when I opened it.

It was filled with puff balls. Fuzzy yellow ducklings called out desperately with their orange bills. Long-necked goslings squawked, and fluffy multicolored chicks peeped. Three odd-looking chicks with an unattractive pimple of skin atop their heads gazed up quietly from the box.

The delivery guy shook his head in disbelief. I could tell he had questions. Were we not in the city? Wasn't downtown Oakland only ten blocks away? Who is this insane woman? Is this even legal? But years of working for the government had, perhaps, deadened his curiosity. He didn't look me in the eye. He didn't make a sound. He just jumped back into his postal jeep and drove away.

Mr. Nguyen giggled. For the last few years he had happily observed—

and participated in—my rural-urban experiments. He knew poultry when he saw it: he had been a farmer in Vietnam before enlisting to help the Americans during the war. "Oh, yes, baby chicks," he said. "Ducks." He pointed a cigarette-stained finger at each species. "Goose." His finger paused at the pimpled heads. He looked at me for a hint.

"Baby turkeys?" I guessed. I had never seen a baby turkey either.

Mr. Nguyen raised his eyebrows.

"Gobble-gobble. Thanksgiving?"

"Oh, yes!" he said, remembering with a smile. Then he grimaced. "My wife make one time."

"Was it good?" I asked. I knew that his wife, Lee, was a vegetarian; she must have made an exception for Thanksgiving.

Mr. Nguyen shook his head vigorously. "No, tough. Too tough. Very bad." I thought he might spit.

I closed the lid, and the peeping stopped. Mr. Nguyen went back into his apartment, returning to the blare of a Vietnamese-language television show.

In the middle of 28th Street, I held the box of poultry and waterfowl. The abandoned ghetto where we lived had a distinct Wild West vibe— gunfights in the middle of the day, a general state of lawlessness, and now this: livestock.

I glanced at the invoice connected to the box: "Murray McMurray Hatchery," it read. "1 Homesteader's Delight." I didn't think about it at the time, but looking back on it, I realize that "Homesteader's Delight" does have a rather ominous ring to it.

Every second-rate city has an identity complex. Oakland is no different. It's always trying to be more arty, more high-tech, more clean than it is able.

O-Town is surrounded by overachievers. The famously liberal (and plush) Berkeley lies to the north. The high-tech mecca of Silicon Valley glimmers to the south. Just eight miles west via the Bay Bridge is San Francisco—so close, but the polar opposite of Oakland. SF is filled with successful, polished people; Oakland is scruffy, loud, unkempt.

I've always chosen uncool places to live. I guess it's because I was born in Idaho, rivaling only Ohio as the most disregarded state in the union. Then I lived in a logging town in Washington State whose big claim to fame was a satanic cult. By the time I moved to Seattle (living in the boring Beacon Hill neighborhood), the uncool, the unsavory, had become my niche. When I went traveling and someone warned me—speaking in low tones, a snarl to her lips—not to go to Croatia or Chiapas or Brooklyn, I tended to add the place to my itinerary immediately.

"Whatever you do, don't go to Oakland," a stocking-cap-clad guy at a Seattle barbecue told me when I confessed that I was going to check out the Bay Area on a long road trip/quest to find a new place to live. I made a mental note to check it out.

Bill and I took three months to explore the candidates. At his insistence, we brought our cat. Bill's a tough-looking guy, with shaggy hair and a strut like he's got two watermelons under his arms. His voice is Tom Waits gravel from years of smoking. He might resemble a Hells Angel, but he's really just a love sponge who spends a great deal of time cuddling with our cat. We hit all the cities we thought we might like to live in: Portland (too perfect). Austin (too in the middle of Texas). New Orleans (too hot). Brooklyn (too little recycling). Philly and Chicago (too cold).

But Oakland—Oakland was just right. The weather was lovely, a never-ending spring. There was recycling and a music scene. But what really drove me and Bill away from the clean and orderly Seattle and into the arms of Oakland was its down-and-out qualities. The faded art deco buildings. The dive bars. Its citizenry, who drove cars as old and beat-up as ours.

Because of inexperience and a housing shortage, Bill and I wound up sharing a ramshackle house in the Oakland hills with a pack of straight-edge vegan anarchists. They wore brown-black clothes, had earth names like Rotten, and liked to play violent computer games in large groups in the common room. Sober.

At first I thought it was cute that anarchists had rules. No alcohol. No dairy products. No meat. Then the paradox started to chafe.

Forced by the strict house regulations, Bill and I would have to rendezvous in our travel-worn van in order to take nips off a contraband bottle of wine, gorge ourselves on banned cheese products, and remember the good old days when we oppressed chickens in our backyard in Seattle. And we plotted our uprising.

One night I unearthed an apartment listing on Craigslist that would set us free. I found it during video game night at the house, surrounded by a pack of anarchists in our living room. While they fired imaginary guns on their computer screens, I clandestinely scanned the ad for the apartment. It was reasonably priced and in downtown Oakland. Feeling subversive, we went for a tour the next day.

The first thing we noticed when we came down from the verdant hills into the flatlands—also known as the lower bottoms—was the dearth of trees. Gray predominated. Bill drove, his coffee brown eyes nervously scanning the scene. We passed one green space huddled under a network of connecting on-ramps. A basketball court, some shrubs. It was called Marcus Garvey Park. No one was there, even on an early summer day.

What was happening was liquor stores. Captain Liquor. Brothers Market. S and N. One after another. The surrounding restaurants were mostly fast-food chains: a Taco Bell, Carl's Jr., Church's Chicken. One variety store caught my eye. Its handmade sign used no words, just images: a pair of dice, socks, eggs, toilet paper. Life's necessities. It reminded me of the little roadside *tiendas* in Mexico. It was the third world embedded in the first.

The houses, though dilapidated, had clearly once been lovely homes: elaborate Victorians next to Spanish Mission bungalows, Craftsman cottages, and vintage brick apartment buildings. They were chipped, charred, unpainted, crumbling. Beautiful neglect.

As we cruised the neighborhood we took stock of our potential neighbors. A man wearing a head scarf was singing as he swept garbage out of the gutter in front of his liquor store. A group of old men sat in lawn chairs in front of their apartment building. A blond woman with scabs on her face limped along the street, pausing to ask for spare change from the young black kids on the corner. The kids wore enormous white T-shirts and saggy pants; they counted their bills and stood in the middle of traffic, waving small plastic bags at prospective customers. Clearly a rough crowd.

All these people out on the street—they were characters I had never met in Seattle, or in our more suburban house in the Oakland hills. I was curious, and yet I had to admit it: they scared me. Could I really live here? Walk around the streets without worrying about getting mugged?

The place was a postcard of urban decay, I thought as we turned down 28th Street. Cheetos bags somersaulted across the road. An eight-story brick building on the corner was entirely abandoned and tattooed with graffiti. Living here would definitely mean getting out of my comfort zone.

We came to a stop in front of a gray 1905 Queen Anne. Like almost every other house in the Bay Area, it had been divided into apartments. The place for rent was the upstairs portion of the duplex. Bill and I surveyed the house. The paint was peeling; a bougainvillea sagged in the side yard. It was a dead-end street, stopping at what was once the grass playground of an elementary school.

Bill pointed out that a dead-end street is a quiet street. He had once lived on one in Orlando and got to know all his neighbors. It made things intimate, he said. Just then, a dazzling woman with cropped platinum hair and platform boots peeked out of her metal warehouse door and beckoned us over to her end of the street.

"My name's Lana," she said. "Anal spelled backward." Bill and I exchanged looks. She stood behind her chain-link fence, a 155-pound mastiff at her side. A robed Buddhist monk emerged from the house next door. He and Lana waved. He disarmed his car alarm—the danger of the 'hood trumps even karma—and drove away. Lana gazed at the retreating car and said, "The old monk used to make me bitter-melon soup when I was sick."

Lana told us, in her high, funny voice, that she had lived on "the 2-8" for fifteen years. "It's not bad now," she assured us. "A few years ago, though, I had people running over my roof, firing machine guns. Now it's like Sesame Street." She shook her head.

Lana then pointed at each of the houses and described its inhabitants: a white family she called the Hillbillies in the teal house, a black mom with two kids in the stucco duplex, an apartment house filled with Vietnamese families who wanted to live near the temple. An abandoned building with a sometime squatter. An empty warehouse that no one knew much about. As we took leave of Lana she invited us to Blue Wednesday, a salon for artists and performers she held every week.

"She seems interesting," I said as we walked back to get our tour of the apartment. Our landlords had arrived in their gold BMW.

"We should move in," Bill said, running his fingers through his shaggy dark hair. He didn't even need to see the apartment.

Our soon-to-be landlords were an African couple with socialist tendencies. They led us upstairs for a tour of the bright little apartment. Hardwood floors. A tile-lined fireplace. A backyard. A living room with a view of a 4,500-square-foot lot filled with four-foot-tall weeds. The landlords didn't know who owned the lot, but they guessed that, whoever they were, they wouldn't mind if we gardened there. We gaped at the enormous space. It had an aspect that would guarantee full sun all day. In Seattle we tended what we thought was a big backyard vegetable garden, but this lot—it was massive by our standards. It sealed the deal.

Bill and I grinned on our way back to our hovel in the hills with the vegan anarchists, still giddy from too much California sunshine and the prospect of a new home.

A few weeks later, when we moved into our new apartment, we discovered that our neighborhood was called GhostTown, for all its long-abandoned businesses, condemned houses, and overgrown lots. The empty lot next to our house was not rare: there was one, sometimes two, on every block. And through the vacant streets rolled GhostTown tumbleweeds: the lost hairpieces of prostitutes. Tumbleweaves.

The day we moved into GhostTown, a man was shot and killed outside a Carl's Jr. restaurant a few blocks away. We drove past the crime scene— yellow caution tape, a white sheet with a pair of bare feet poking out. We heard on the radio that Oakland had been named number one—it had the highest murder rate in the country. When we drove by later, the body was gone and the business of selling hamburgers and soda had resumed. That night, the not-so-distant crack of gunfire kept me up.

Because of the violence, the neighborhood had a whiff of anarchy—real anarchy, not the theoretical world of my former roommates. In the flatlands, whole neighborhoods were left with the task of sorting out their problems. Except in the case of murder, the Oakland police rarely got involved. In this laissez-faire environment, I would discover as I spent more time in Ghost-Town, anything went. Spanish-speaking soccer players hosted ad hoc tournaments in the abandoned playfield. Teenagers sold bags of marijuana on the corners. The Buddhist monks made enormous vats of rice on the city sidewalk. Bill eventually began to convert our friends' cars to run on vegetable oil. And I started squat gardening on land I didn't own.

As I fiddled with the door to our apartment, the new box of fowl tucked under my arm, I recognized that I was descending deeper into the realm of the underground economy. Now that I had been in California for a few years, I felt ready for what seemed like the next logical progression, something I had never dared in the soggy Northwest.

Meat birds.

I felt a bit nuts, yes, but I also felt great. People move to California to reinvent themselves. They give themselves new names. They go to yoga. Pretty soon they take up surfing. Or Thai kickboxing. Or astral healing. Or witch camp. It's true what they say: California, the land of fruits and nuts.

In Northern California one is encouraged to raise his freak flag proudly and often. In Seattle my mostly hidden freak flag had been being a backyard chicken owner, beekeeper, and vegetable gardener. I got off on raising my own food. Not only was it more delicious and fresh; it was also essentially free.

Now I was taking it to the next level. Some might say I had been swept up by the Bay Area's mantra, repeated ad nauseam, to eat fresh, local, free-range critters. At farmer's markets here—and there is one every day— it isn't uncommon to overhear farmers chatting with consumers about how the steer from which their steaks were "harvested" had been fed, where their stewing hens ranged, and the view from the sheep pen that housed the lamb that was now ground up and laid out on a table decorated with nasturtium blossoms. Prices correspond with the quality of the meat, and Alice Waters assures us that only the best ingredients will make the best meals. But as a poor scrounger with three low-paying jobs and no health insurance, I usually couldn't afford the good stuff.

Since I liked eating quality meat and have always had more skill than money, I decided to take matters into my own hands. One night, after living in our GhostTown apartment for a few years, I clicked my mouse over various meat-bird packages offered by the Murray McMurray Hatchery Web site. Murray McMurray sold day-old ducks, quail, pheasants, turkeys, and geese through the mail. They also sold bargain-priced combinations: the Barnyard Combo, the Fancy Duck Package, the Turkey Assortment.

These packages, I had thought, might offer a way to eat quality meat without breaking the bank. But I had never killed anything before. Blithely ignoring this minor detail, I settled on the Homesteader's Delight: two turkeys, ten chickens, two geese, and two ducks for $42.

I bought my poultry package with a click of the mouse and paid for it with a credit card. It was only after the post office delivered the box that I realized one can't just buy a farm animal like a book or CD. What I now held in my hands was going to involve a hell of a lot of hard work.

My first task was to install the birds in a brooder, a warm place where they could live without fear of catching cold or encountering predators. I carried the box o' birds upstairs and set it next to the brooder I had hastily built the night before. "Built" might be a strong word—my brooder was a cardboard box lined with shredded paper, with a heat lamp suspended above it and a homemade waterer inside.

The hatchery advised that the chicks would be thirsty from their twenty-four-hour journey in a box. So the first order of the day was to dip the birds' beaks into a dish of water and teach them to drink on their own.

I picked up my first victim, a little yellow chick covered in a soft, downy fuzz, and held her tiny pink beak up to the homemade waterer. It consisted of a mason jar with tiny holes drilled into the lid; when the jar was turned upside down into a shallow dish, capillary action allowed only a bit of water to dribble out and pool in the dish. Amazingly, the chick knew just what to do. She sipped up a beakful of water, then tilted her head back to swallow. The mason-jar waterer glugged, and more water seeped out.

I released her into the cardboard-box brooder, and she wandered over for another sip of water. Then she realized she was alone. She peeped and stumbled around the shredded newspaper looking for her companions. The fowl still in the postal box, strangely silent since I'd placed it on the living room floor, suddenly went wild when they heard her peeps.

So I reached into the box for another chick and worked quickly. Without fail, each victim peeped in distress. The others then chirped in solidarity. All ten finally installed, the chicks quieted down. Exhausted from their journey and my manhandling, they mounded into a fluffy pile under the circle of warm light and took a nap.

Bill stumbled out of our bedroom wearing his boxer shorts, his hair mussed. Not a morning person, he glanced at the baby birds like they were a dream, then headed for the bathroom.

While the chicks slept, I had to educate the dim little turkey poults. They looked like the chicks but with bigger bones and that strange pucker of skin on top of their heads, which I later learned would develop into a turkey part called the snood. Their demeanor was reminiscent of chicks that had done too much acid.

It took the first turkey poult three firm dunkings before it got the hang of drinking water. The poult resisted when I put its beak into the dish, craning its head away, struggling in my hand like a hellcat. Finally, exhausted from struggling, its head went lax and drooped until it dropped into the water dish, where it discovered—surprise!—water, and drank greedily. The other two (the hatchery had sent me an extra poult and an extra duckling, probably as insurance against death by mail) were no different. After I released them, the poults poked around the brooder, gentle and cautious. Eventually they waddled over and joined the puff pile of chicks.

The downy, almost weightless ducklings and goslings drank deeply, using their bills to slurp up large amounts of water. When I set them into the brooder, they waded their big orange feet into the water dish and splashed around. Water hit the side of the box and splattered the sleeping chicks, who awoke and began to peep in protest. Sensing that this might be a disastrous species intersection, I lugged out an aluminum washtub and set up a separate brooder with extra water, a towel, and a bright warm light for the waterfowl.

The baby birds were home, warm and safe. The chicks scratched at their yellow feed just like our big chickens out back did. Sometimes they would stop midscratch and, feeling the warmth of the brooder light, fall asleep standing up. The puffy gray goslings curled their necks around the yellow sleeping ducklings. A Hallmark card had exploded in my living room.

I called my mom. A brooder box full of fowl was something that woman could appreciate. She had once been a hippie homesteader in Idaho.

"Listen to this," I said, and held the phone near the brooder box. A hundred little peeps.

"Oh my god," she said.

"Three turkeys, three ducks, two geese, and ten chickens," I crowed. I watched the chicks and poults moving around the brooder—pooping, scratching, pooping, pecking, pooping.

"Turkeys! Do you remember Tommy Turkey?" she said.

I didn't, but the photo in our family album had stuck with me: my older sister, Riana, in a saggy cloth diaper being chased by the advancing figure of a giant white turkey. Tommy. My mom told us about Tommy every time we got out the old photo album from the ranch days.

"Well, he was mean as hell, and he would chase you guys. . . ."

I looked out the window while my mom described the smokehouse she and my dad had built. Bill had made it downstairs, where he was out front tinkering with our car. His legs peeked out from underneath our dilapidated Mercedes as he rolled around amid the street's numerous Swisher Sweet cigar butts. I had warned him about my meat-bird purchase, and he had been excited about the prospect of homegrown meat, but now that he saw the baby birds—fragile, tiny—he seemed a bit skeptical.

Tommy grew to be an enormous size, my mom said, and as back-to-the-land hippies, she and my dad had been very pleased. They didn't encounter any predator problems that year, and butchering him was a cinch. But disaster did hit: the smokehouse burned to the ground while they were smoking the turkey.

"Oh, no," I groaned.

"Life was like that," she said glumly. I felt sorry for her. My mom's stories usually involve some heroic hippie farm action. I hadn't heard this part of the story before, but I knew bad things had happened. My parents' marriage had dissolved on the ranch in Idaho, after all—my dad too much the mountain man, an uncompromising nonconformist; my mom isolated and bored.

Her voice brightened. "Even though the smokehouse burned down, we did manage to salvage the turkey."

"What do you mean?" I asked.

"We dug through the charred wood, and there it was, a perfectly cooked turkey. I brushed off all the cinders and served him for dinner." She paused and smacked her lips, a noise that was repellent to me as a teenager but now filled me with hope. "It was the best turkey I've ever had," she declared. We said our goodbyes, and I hung up the phone.

I glanced into the cozy chick brooder. The chicks slept on their mattress of shredded pages from the *New York Times*. Their fuzzy bodies slumbered on snatches of color ads for watches, a stern op-ed about pollution in China, the eyebrows of a politician. I had to remind myself that though they were cute, these baby birds would eventually become my dinner. Thanksgiving, in particular, was going to be intense. I imagined the killing scene: a butcher block, an ax, three giant Tommy turkeys I had known since poulthood. I wasn't sure if I could bring myself to do it.

But the conversation with my mom left me emboldened for my foray into killing and eating animals I had raised myself—this urge was clearly part of my cultural DNA. I wondered if this would prove that I could have it both ways: to sop up the cultural delights of the city while simultaneously raising my own food. In retrospect, though, I wonder why I thought my experience would be any less disastrous than my parents'.

The next day, following the suggestions of a homesteading book from the 1970s, I swabbed the baby birds' butts with Q-tips. The long flight in a box can cause digestion problems for the chicks—namely, pasted vents. Which is a fancy way of saying blocked buttholes. So I dutifully wetted them down, plucked dried matter from their bottoms, and felt terrible when I had to tug off whole chunks of downy feathers. I wasn't satisfied until all their parts looked pink and healthy.

After morning chicken-butt detail, I sat in my kitchen and surveyed our squat garden. All the east-facing windows of our apartment overlook the lot, which after the past few years had been transformed into a vegetable and fruit-tree garden. I could see that the collards were getting large and that the spring's lettuce harvest promised to be a good one. Even from inside, I could see some mildew forming on the pea vines.

It was going to be a remarkable year; I could sense it. If my life in Oakland was a developing embryo, with this meat-bird addition, it was as if a fishlike creature had suddenly sprouted wings.

CHAPTER TWO

The garden—squat, verdant—started small, and in fits and starts. When we moved into the apartment on 28th Street, Bill and I halfheartedly painted our walls and hung some curtains. Then we started in on the lot, the real reason we had picked the apartment. We spent days hacking back the four-foot-tall weeds that had taken over, revealing a cracked concrete foundation where a house had once been and a large, circular patch of dirt.

Before we planted anything, just to be safe, I had the soil tested by a friend with an environmental-services lab. The lot being so close to the freeway, the lead from years of gas exhaust might have drifted down. Or other toxins from the house could have leached into the ground. Our friend called with good news: the soil was, miraculously, heavy-metal free.

The day after getting the green light, I stood in the lot, trying to get my nerve up to garden. I was having a tough time getting used to the idea of cultivating land that was not my own. Cutting down weeds was one thing, but planting seeds?

As I stood immobilized, our new neighbor Lana, dressed in patent-leather combat boots and a miniskirt, stomped into the now cleared-out lot. Everything about Lana was theatrical: expressive hazel eyes, a gap-toothed smile, and a platinum crew cut that matched the color of the fallen weeds. "Look at this!" she shouted. She held two shovels in her strong arms.

"There's a patch of dirt here," I said, showing her where the yard to the house may have been. Even though she had never gardened before, Lana set to work with enthusiasm. I was glad to follow along. Her dog, Oscar, sniffed at the piles of dead weeds and did a little digging himself. Gentle for such a large dog. We dug out a small area, unearthing rusty toy cars, submerged bricks, and glass bottles.

"Do you remember He-Man?" Lana asked, wagging an action figure in front of my face. The toy had big muscles and a bowl haircut but had been drained entirely of any color. Lana put He-Man and all the other toys she found in a small section of the garden.

I had been reading a book from the library called *Gaia's Garden*. It was a permaculture guide to gardening that promised I could create an easy-to-maintain, no-work food forest if I just followed the instructions. Plans were included for something called a keyhole garden: a series of pathways cut into a circular bed—which is what we happened to have. So Lana and I set to work.

After a few hours, we had finally cleared enough space to plant some seeds. Just before I ripped open the packet of corn, Lana alerted me to a problem. "Ah, Novella," she said, wiping sweat from her brow, "is this supposed to be in the shape of . . . ?" She trailed off.

"What?" I said, still unsure about Lana. My fingernails were caked with soil.

Lana laughed a big, booming cackle. "It's like a crop circle," she said.

Then I saw what she was talking about: following the ditzy plan, I hadn't realized that we had built a garden in the shape of an enormous peace sign, which, viewed from the heavens, might be some kind of hippie signal to the aliens.

I was horrified.

"Oh my god," I said. "Glad you pointed that out." I erased one of the pathways with a few shovelfuls of dirt so that the peace sign became more like a Mercedes-Benz emblem. Leave the peace sign to followers of the Dead and wearers of tie dye, to my hippie parents' generation. Even though Lana and I were vegetable gardening, we wanted to be clear with each other—especially so early in our friendship—that we were *not,* in fact, hippies.

As Lana watched I tucked a few kernels of corn into the dark clay soil, feeling strangely shy. I wasn't used to being a squat-garden rebel, though on an intellectual level the idea of squatting, of taking possession of something unused and living rent free, had always held a certain appeal for me. In college I read about the Diggers (also called the True Levellers) in seventeenth-century England, who squatted in houses and planted vegetables on public land. In 1649, a scroungy group of men gathered at a small town southwest of London to plant corn and wheat on the commons. In the declaration they submitted—mostly a bunch of Bible talk—explaining why they were "beginning to plant and manure the waste land of George-Hill," they expressed their belief that the earth was "a Common Treasury for All, both Rich and Poor, That every one that is born in the Land, may be fed by the Earth his Mother that brought him forth." Almost 350 years later, the idea of planting food crops in common areas still makes a great deal of sense.

In America, squatting dates back to the very beginning of white settlement. Seeking religious freedom, the Puritans, let's face it, squatted on Indian land. Pioneers in the 1800s continued the process by squatting on more Indian land during the Westward Expansion. In the 1980s, the tradition continued when abandoned buildings in New York City were taken over by squatters like crazy. In 1995, I was befriended by a squatter who lived in a building on Avenue B in New York. Though I was game to join and move in, in the end I wasn't deemed punk-rock enough—maybe because I didn't have tattoos or spikes on my clothing.

And then there was the famous bean sower Henry David Thoreau. He didn't own that land near Walden Pond or even rent it. "I enhanced the value of the land by squatting on it," he wrote in *Walden*.

That was my plan, too. I took a deep breath and plunged the little yellow seeds into the ground that was not mine. I snaked the hose around from our backyard and sprinkled water onto the bare patch of soil. Lana and I stood and watched the water soak in. What I was doing reminded me a little of shoplifting, except instead of taking, I was leaving something. But I was worried. Couldn't these plants be used as evidence against me?

Within a few weeks of that first sowing, I grew more accustomed to the idea that the lot was temporarily mine. I transplanted a few tomato and basil starts. Sowed the lettuce seeds I had carried, pioneer-like, with me from Seattle. Planted a few cucumber seeds.

It wasn't long before the vegetables grew and thrived in the blazing, all-day sun. Their success had nothing to do with my skill as a gardener—in California, it's just-add-water gardening. The cukes scrambled; the corn lumbered toward the sky.

Lana told me that the lot had been, over the fifteen years she had lived on 28th Street, first a parking lot for the monks; then a storage space for a construction company, replete with shipping containers and Bobcats; and finally, for the last five years, the weed-filled, garbage-strewn dump we found when we moved in. The garden, by comparison, did seem like an improvement. But in the back of my mind I wondered what the owner of the property would think of all the new plants.

Upstairs in the kitchen a few days after the chicks arrived, I poured myself a cup of coffee, then went to the back stairs to throw some day-old bread to the big backyard chickens (a Buff Orpington, a Black Australorp, and two Red Stars), who had come, full-grown, from a nearby feed store.

A word about my backyard: Don't entertain some bucolic fantasy. In the middle of it, Mrs. Nguyen's exercise bike sat on a bare patch of dirt. The landlord had installed a rusty metal shed at the foot of the stairs a few years earlier, and it now held all the things we and the Nguyens wished we could get rid of. A shattered mirror lay between the fence and the shed.

In the very back of the yard was the chicken coop Bill had built from pallets in what had been a large dog run made with sturdy chain-link fencing now overgrown with weeds and volunteer trees. Abutting this chicken area was an auto-repair shop/junkyard, which hosted two dogs: a pale brown pit bull and a dark-eyed Rottweiler mix. A forklift often zoomed around the repair shop, dodging rusting transmissions and barrels of god knows what. A little beyond the auto shop, you could see downtown Oakland's nondescript skyline. Not exactly a country idyll.

After feeding the backyard hens and checking (again) on the baby chicks in their brooders, I sat down to read the paper. After a few minutes I looked up and noticed a man entering the garden, wearing a black skullcap and a leather jacket. He wandered over to one of the beds and tugged on the green top of a carrot. An orange root appeared, streaked with dirt. The carrot, I could see, was small but edible. Throwing my paper down, I rushed to the garden.

People from the neighborhood harvesting food from my garden is a common sight. There's Lou, a stooped man who helps himself to the lush crop of greens in the winter; a mute lady who carries a plastic sack into the garden and doesn't stop harvesting lettuce until that bag is swollen—or until I open the window and call down at her, "OK! That's enough! Leave some for everyone else!"

Some of the harvesters are annoying. One year, an unidentified person stopped by for what he thought was onions and picked some of my young garlic instead, then abandoned the small bulbs on the ground. In response, I made a little handwritten sign that said GARLIC, NOT ONIONS. READY IN JULY! and another, near a collard-greens patch, saying DON'T PICK ALL THE LEAVES OFF THE PLANT. These signs aren't necessarily effective. They just fade and get buried by a pile of wood chips in the fall. But I feel the need to instruct nonetheless.

A simple solution would be to snap a padlock on the gate. Then again, I'm a trespasser myself—I don't lease or rent the verdant lot, so I'd feel like a hypocrite telling others to stay away. But I did at least get approval for the garden project from the owner of the property, a man named Jack Chan. I met him when our first tomato ripened in the lot that first year. A wizened Chinese man walked into the garden. I could tell he was the property owner by the way he walked past the gate and looked at the plants—quizzically, as if they were a magic trick he couldn't quite figure out. My heart pounding, I went down to talk to him. "Garden OK," he said after we made introductions. Then he pointed to a few nongarden items that had made it onto the lot, like some old doors and a biodiesel reactor Bill had built. "Only garden." I nodded, and that was the end of our exchange.

If I was trying to be Thoreau, I liked to think of Chan as a modern-day version of Ralph Waldo Emerson, the owner of Walden Pond and its surrounding fields. My fellow squatter Thoreau *did* have permission from the landowner, but he still liked to call what he was doing—just as I did—squatting.

Once I got Jack Chan's terse seal of approval, I began enhancing the land big-time. The next year the whole lot sprawled with giant orange Rouge Vif d'Estampes pumpkins. I had a customer-service job at a plant nursery and got discounts on fruit trees, so in went an apple tree, a pineapple guava, a lemon, a fig, and an orange tree. I bought bees to pollinate the ancient plum tree that stood guard in the lot. I acquired my four egg-laying chickens and grew dinosaur kale, a crinkly, dark green variety, especially for them. I planted carrots, which this guy was now harvesting.

I pushed open the gate and called out a hello to the carrot picker. He waved, holding the carrots in his hand. I know about the pleasure of pulling up root vegetables. They are solvable mysteries. I once pulled up a carrot unlike any I had seen before. It was a deep purple variety called Dragon, and it had wound itself around a regular orange carrot, so they looked like a gaudy strand of DNA.

Just as I was about to tell the carrot picker that he should come back to harvest the carrots when they were bigger, he said, "This place reminds me of my grandma." His eyes filled with tears. "Everything's so growing," he said.

In our neighborhood, there was some greenery, mostly in the form of weeds. But when you walked through the gates into what I had started calling the GhostTown Garden, it was like walking into a different world. There was a lime tree near the fence, sending out a perfume of citrus blossoms from its dark green leaves. Stalks of salvia and mint, artemesia and penstemon. The thistlelike leaves of artichokes glowed silver. Strawberry runners snaked underneath raspberry canes. Beds bristled with rows of fava beans, whose pea-like blossoms attracted chubby black bumblebees to their plunder. An apple tree sent out girlish pink blossoms. A passionfruit vine curled and weaved through the chain-link fence that surrounded the garden.

I restrained myself from hugging the carrot picker for feeling exactly as I did about the garden, but I did get a little misty. I wanted to grab this man's arm and give him a tour, show him what's edible, what will be at its peak next week, which part of the mint to snip off for tea. Pull up a few of the French breakfast radishes. Explain that carrots are native to Afghanistan and used to be tough and yellow before the orange-loving Dutch got hold of them. Then I'd take him to the backyard and show him my four prized chickens, their straw-lined nesting boxes, the four eggs from that day— brown and still warm. Maybe I'd take him upstairs to admire the brooder box of baby chicks, the waterfowl, the turkey poults.

This, I wanted to tell him, is your birthright, too. Your grandmother, like mine, grew her own tomatoes, killed her own chickens, and felt a true connection to her food. Just because we live in the city, we don't have to give that up.

But then I remembered that most people in our neighborhood have other things on their minds than growing local organic food and starting a revival. I know because when we first started, I knocked on doors to get other people involved with the work in the garden, wide-eyed and bushy-tailed about growing vegetables. I got stock responses: "I don't have time" or "I buy my food at the grocery store."

So I was just glad to be a reminder for the carrot picker. I muttered that he should come by and harvest food anytime. He smiled, exposing a glimmering grill of gold.

"Hey, my garden's your garden," I said, and patted him on the back. He left with a small bunch of carrots stashed in the pocket of his leather jacket. I never saw him again.

I picked a handful of kale leaves for the chicks and went upstairs. I cut them into a chiffonade and watched the chaos unfold when I tossed the green strands into the brooder boxes. A pure yellow chick grabbed the biggest piece of green and carried it around while making an urgent peeping noise. A few others gave her chase until she dropped it. Then another chick picked it up and ran and peeped. It was all about the having, not the eating. This went on for quite some time until the turkey poults, who had been watching the proceedings heavy-lidded, snatched away the greens from the chickens and gobbled them up. The ducklings and goslings, in their own brooder, inhaled the kale without hesitation. The squat garden would feed my meat birds, too.

❁

CHAPTER THREE

B obby!" Bill and I yelled toward the end of the dark street, not sure which car he was sleeping in that night. "We need your help!" It was Sunday, late, and we were on a gardening mission.

The door to a Chevy Colt without wheels creaked open. Bobby came out, sighing, and seemed annoyed, but I think he liked that we needed him. Two people weren't enough to push one of Bill's project cars out of the way.

Arms sprawled out, I heaved against the back of the marooned red Mercedes. Bill steered and pushed up front. Then Bobby pressed his back against the trunk of the car and leaned his weight in. That's all it took.

"You gotta use your leg muscles," he said, casually walking against the car as it rolled. It wasn't just about technique, though. Bobby was astonishingly strong. I once saw him lift the transmission of a Ford truck from the ground to his shoulder and carry it for a city block, seemingly without effort. Bobby wouldn't take money for his help; he would mutter, "Life of the Hebrews," when we offered.

We first encountered Bobby during a 2-8 squabble. We had been living in our apartment for more than a year, long enough to have permanently tuned out the sound of the traffic from the nearby highway. The sirens no longer fazed our cat. We were making friends and had sussed out the best late-night Chinese-food joints. As former Seattleites, we were amazed by and thankful for spring's sunny arrival in February.

We had also begun collecting stuff, one of our favorite hobbies. Our tables and desks were scrounged from street corners in Berkeley and Oakland. Most of our dishware came from free boxes set out on corners. There was something captivating about making something useful again—resurrecting the abandoned.

So we were in our element in Oakland, with its mammoth piles of junk placed on curbs, clutter dropped under overpasses and, sometimes, in the middle of the street. The junk piles became so bad that at one point there were billboard ads urging people to DUMP BOYFRIENDS, NOT APPLIANCES. It was a strange campaign—stranger when half the lights on the billboard went out, leaving only the illuminated command DUMP BOYFRIENDS.

GhostTown in particular hosted some huge piles. Most appeared near the beginning of the month, when rent was due. Sagging couches, black lacquer nightstands, mattresses, all stacked up on the corner—a surprisingly intimate life-size scrapbook of someone's existence. Eventually city maintenance workers would mark the pile with a blast of orange spray paint and haul it away. If they weren't fast about it, the pile would become a sprawling, multihouse, multiblock affair.

These piles were sort of like an ecosystem—a complete community of living organisms and the nonliving materials of their surroundings. Some individuals added to the growing mass, bottom-feeders harvested from the pile, and sometimes the items broke down into dust of their own accord. In this ecosystem, Bill and I played the role of bottom-feeders.

One night, Bill spotted a blue bicycle in a pile of junk on the corner of our street. It was a vintage affair, with three speeds and only slightly messed-up back brakes. He carried it upstairs and spent the evening tinkering with it.

Out for a victory ride the next morning, Bill heard, "That's my bike!" It's a common refrain on the streets off Martin Luther King Jr. Way. But the man who yelled it, standing on the porch of his warrenlike apartment building, didn't mean it in jest. He walked toward Bill. A not quite elderly black man, only slightly grizzled, missing most of his teeth. Name: Bobby.

Seeing that this might get ugly, I walked downstairs to intervene: "Sir, you know the rules around here. If it's in a pile, it's up for grabs." Bobby didn't say anything. It was apparently a man thing, between Bill and Bobby.

"I fixed it," Bill pleaded.

Bobby put his hand on the handlebars. "Uh-uh," he said. Who is this asshole? I thought.

But Bill wasn't going to fight about it. I could tell he had second thoughts. He looked down, then up. "I'll give you $20," Bill said, excited at the possibility of a solution. At that, Bobby took his hand off the handlebars and beamed. Bill passed him the bill. The matter concluded, Bill rode off on his bike. A beautiful friendship had—yes—blossomed.

Now Bobby helped us do things like move cars around.

"OK, stop," Bill yelled to me and Bobby. We let up. Bill jumped into the Mercedes and braked—Bobby and I had gotten a bit too zealous in our pushing. The red Mercedes barely avoided smashing into our neighbor's Honda.

Bill swung open the gates to the garden. The F-250's wheels spun as I jumped the curb and backed into the lot. Our friend Willy had loaned us his truck for the weekend, so Bill and I had spent Saturday and Sunday making runs to a horse stable fifteen miles away, up in the hills. The free rotted horse manure had been our ticket to gardening success.

Since most of the lot was under a one-foot layer of concrete, Bill came up with the idea of building raised beds. Most vegetables don't require more than a few feet of topsoil, so it's entirely possible to grow plants in large containers. We made open-topped boxes, filled them with composted horse manure, and planted the majority of our herbs and vegetables in them.

Bill, the ultimate bottom-feeder, wouldn't even buy wood to build the beds. In the early days of the lot, Bill would return home, his dark hair mussed, a crooked smile of delight on his face, the borrowed truck filled with sheets of plywood and odd chunks of lumber found along the streets of GhostTown. In the massive abandoned piles, Bill found garden-building materials.

We learned that four chunks of a two-by-four made corners for boxes and connected everything. I got good with an electric handsaw. The only things we bought were screws and an electric drill.

Now in our third spring of gardening, we were still building new beds and topping off the existing ones. Every year, gripped by the fever of spring gardening, our mantra was always "More manure, we need more."

A series of raised beds, like coffins, scattered the lot. That weekend, we had topped off the three existing beds and begun dumping the rest into new, yawning, empty boxes. I parked near the biggest one, cut the engine, and jumped into the truck bed.

Bill was already there, sinking the shovel into the crumbly black gold. We had only one shovel, so I squatted down, facing away from the bed, and used my hands to bulldoze the manure between my knees over the edge of the truck. Bobby watched us unload the soil with a mix of curiosity and disgust.

In February, Bobby had been kicked out of his apartment. This is why he was living in a car parked in front of his old house. We never asked why he had been evicted, though he was seeming less and less lucid, so we had our suspicions. He had become the unofficial security guard of the 2-8.

That he lived in a series of cars wasn't the kind of thing that raised eyebrows on our street. One neighbor, after arguments with his girlfriend, regularly retires to his car—a cream-colored BMW with the windows knocked out and replaced with Mexican-soda cardboard boxes. When we see him shouldering a bag of clothes in one hand, headed to the BMW with a defeated slouch, Bill and I look at each other and say, "Someone's in the doghouse." So Bobby's new home seemed perfectly acceptable.

Besides becoming a squatter, Bobby became a farmer, too—only his crops were cans and metal. He hauled them via a shopping cart to the recycling center a few blocks away. Like a pack rat, he also collected other items: backpacks, light fixtures, exercise equipment. Anything that once had value (but now was stained and smelled weird) Bobby would take home with him. And home was that wheelless Chevy Colt.

"So that's horse poop?" he asked while we unloaded the rest of the manure.

"Yes," I panted, picking up a bucket and scooping out the corners of the truck bed. When I looked up into our apartment, it was dark except for a warm yellow glow in the living room—the brooder box. The chicks were getting bigger—and louder. They often woke me at dawn with their squabbling.

"And you grow vegetables in it." Bobby was wearing a pair of antennas he made out of a girl's headband and some tinfoil.

"Yes, it's really composted down, though," I assured Bobby. I stood up to stretch my back but found I couldn't stand up completely. I hunched over, my shoulders caved in, and gasped, "So there aren't any bad bacteria or whatever."

"We used chicken droppings," Bobby said. "*Whoo,* that stuff stunk. Now, this isn't too bad." He took a pinch of the manure and sniffed. Bobby had come from Arkansas as a young man. Many of the black people living in Oakland came from families who had migrated from the South in the 1920s to work as longshoremen for the port, as porters for the railroad, or in manufacturing jobs. Back then, Oakland was known as the Detroit of the West. In the 1940s, in what some historians call the second gold rush, manufacturing and military jobs attracted more immigrants from the South, and the black population grew by 227 percent. Oakland, once a monoculture of whiteness, became diverse when people like Bobby's parents moved in.

Bill and I surveyed our progress in unloading the horse manure. The truck bed was empty. The raised bed was . . . half empty. I stared at it with contempt. I was exhausted, but this was our last chance to use our friend's truck. We would have to make another run.

"Can you make sure no one parks here?" Bill asked Bobby. We needed the area in front of the lot clear so we could unload our next load of horseshit. Bobby nodded and went to get a shopping cart to block the parking spot. He waved at our truck as we drove away, back to the hills, back to the stables.

We had to cross the county line to get our horse poo. Oakland's county, Alameda, gave way to Contra Costa County, land of rolling hills, working cattle ranches, and more recently rich folks with McMansions. Lucky for us, rich people like horses. And horses make a lot of manure. Which piles up and composts away until an enterprising gardener arrives and offers to take away this jackpot of tilth and nutrients.

The horses whinnied when they heard us drive up. I backed the truck as close as possible to the mother lode: a massive mound of composting manure

the size of a small barn. The smell—horse sweat, dirt, grass, and that unmistakable odor of cellulose breaking down—was heavenly. It reminded me of
growing up on my parents' property in Idaho. Two of my favorite family
photos are one of my father astride a gorgeous pinto in a snowy field and
another of me riding a brown pony.

I was only four years old when my parents' life on the ranch finally unrav-
eled and my mom, my sister, and I moved to town. I had my first existential
crisis when I realized that it was not possible to have a pony in the city. I still
remember standing in my bedroom, looking out my window, and feeling
the utter horror and emptiness of my horseless life in town. Eventually I got
some unicorn posters, and all was healed. Or maybe not all, because I still
feel a prickle of almost religious ecstasy at the smell of horseshit.

Our buckets clattering, Bill and I marched up to the edge of the pile. My
method was to cradle a bucket in my arms and scrape the side of the manure
hill until a mini avalanche filled the bucket. Bill used a shovel to scoop
from the very bottom of the pile. Red worms came along with the black
dirt, which was warm to the touch. It steamed a little in the chilly night air.
Bucket after bucket until we filled the back of the truck. It was our third trip
of the day, it was night, and our arms were aching from the schlepping.

We paused in our bucket filling and noticed the silence. No highway
noise, no car alarms or ambulances. The hills unfolded off to the east, little
farms marked by a light or two. We were truly in the country.

Driving away from the stables, the truck's suspension nearly buck-
ling under the load, I looked back at the massive hill of manure. It looked
untouched.

"Man, Willy's going to be pissed when he finds out how much manure
we loaded into this thing!" I said.

"Let's not tell him," Bill suggested.

Farther down the road, a fog had rolled in and enveloped the hills that
looked out over the East Bay.

"Well, we'll just give him some tomatoes or—"

"Look out!" Bill cried and grabbed the truck's Oh Shit handle.

We had almost veered over a cliff. I'm a horrible driver, once almost launching us into the Pacific Ocean while driving along Highway 1. I braked and slowed down and started to really concentrate on the road.

"God," Bill said.

"Sorry," I muttered, and we fell silent as the truck rattled down the road. With the low visibility, everything suddenly felt treacherous. A strange loneliness filled my heart, and I thought of my mother.

The road to our ranch in Idaho had been similarly treacherous, and I remembered her story about the day I was born. It was late December, and my parents had hoped to win the New Year Baby contest put on by Les Schwab Tires in Orofino, Idaho. The parents with the first baby born on January 1, 1973, would win a pair of tires and a side of beef. My parents thought they had timed it perfectly, but I was a restless little baby and emerged instead on the snowy night of December 30.

When my mom tells the story of my birth, which has become part of the popular lore of my family, she paints a colorful picture. There was three feet of snow on the ground, and the truck barely avoided sliding off the steep ravine near the ranch. Then the truck threw a rod, destroying the engine, so they had to hitchhike to the hospital. She always tells the story with a smile, as if it had all been a great deal of fun. But now that I'm an adult, when I hear her story, it sounds dangerous, frightening, cold—distinctly unfun.

I cracked open the window of the truck to stop the condensation on the windshield and braked slowly around a hairpin turn.

The country had taken a toll on my mom. She was lonely up there on the ranch. My dad, who eventually went semiferal, would often go on weeks-long hunting trips, leaving my mom to tend to the ranch duties: milking the cow, watering the garden, and locking the duck pen at night. She missed her friends, her exciting life when she had attended be-ins in Golden Gate Park, danced at rock shows, and traveled the world.

I still regard the country as a place of isolation, full of beauty—maybe—but mostly loneliness. So when friends plan their escape to the country (after they save enough money to buy rural property), where they imagine they'll split wood, milk goats, and become one with nature, I shake my head. Don't we ever learn anything from the past? And that's probably why I avoided rural places and chose to live in the city—but, of course, my modified, farm-animal-populated version of the city.

The fog broke once we hit the highway. Propelled by the weight of the manure, we swooped down the concrete mainline of Highway 24 back into Oakland with a fine dusting of horseshit trailing behind us. My melancholy mood was replaced by a wave of love toward my adopted city. With its late-night newsstands and rowdy bars, a city meant I would never be lonely.

When we turned down our street, Bobby was there, guarding the gates.

Bill and I met on an elevator, fell in love because of cats, and lasted because of bees.

In 1997, I was headed to a class to show David Attenborough's *The Private Life of Plants* to a group of Ecology 101 students. While finishing up my degrees in English and biology at the University of Washington, I worked as a projectionist, paid $3.85 an hour to hit PLAY on a VCR and then sit back in the AV booth and do my homework.

Classroom Support Services, my employer, had recently hired a skinny new guy who wore an ugly red wool hat and a too-short sweatshirt. He was in the elevator when I got on, and he scrunched up against the wall and seemed extremely nervous. I gave him a smile, and he returned it with a half wave. I like nervous people, because they make me feel confident. He was cuter than I initially thought, with olive skin and warm brown eyes.

At my floor, I stepped off the elevator.

"Um, excuse me," the man stammered. He had cotton balls stuffed in his ears. Later I would find out he had problems with his ears, especially in the cold wet of Seattle. The cotton balls kept out the elements, as did the red hat.

He handed me a folded sheet of yellow paper. I glanced at it—*The Speckled Pig Zine,* it said. The doors closed, and I walked to my class.

A few minutes later, while David Attenborough's British-accented voice filled the auditorium, I looked through the zine in the booth. Some funny poems, a story about a lost dog, and a questionnaire mostly about cats. (You see a cat. Do you, a. kiss its head? b. kiss its paws? c. kiss it on the lips?) I find men who have felines impossibly sexy.

On our first date, he gave me a ridiculous pair of rabbit-fur gloves he had found on the bus. They were turquoise with a white fur lining. I loved them. We walked around to various bookstores. It was cold, and he took my arm and leaned in to smell my hair. Later I met the cats, Speck and Sparkles, and saw Bill's tiny studio apartment.

Bill had grown up in Indiana and Florida. His mom was from West Virginia, a strapping farm girl with ten brothers and sisters who helped her mom raise chickens and pigs on their little five-acre farm. She and Bill's dad had gone into construction and built fancy houses in Florida. Bill hated Florida and had recently moved to the other end of the country.

We moved in together after our second date. We settled into a rambling house on Seattle's Beacon Hill that became known as the Hen House.

For my twenty-fifth birthday, Bill loaded me into the car and we drove toward Mount Rainier. We pulled into a U-cut Christmas-tree farm and gift shop, and I wondered why he thought I would want a Christmas tree for my birthday. Plus, it was December 30—was he not only totally off base but also incredibly cheap? Maybe I had really misjudged this guy, I thought, looking at a beeswax candle in the shape of a gnome in the store. While I pondered my bad-gift future with Bill—snow globes, kitten-statue doorstops, balloons that read I WUV U—I weighed the merits of our relationship. Great pillow talk. A love of reading. A similar sense of what is funny. Gift hell would have to be a concession.

After he wandered around the rustic cabinlike store, which smelled like cinnamon sticks and pine needles, Bill stopped in front of me. "Novella," he started in his soft but gravelly voice. He scooted me closer to some pine-colored boxes stacked up near the door. "I'm giving you a beekeeping kit for your birthday."

He pointed at the hive boxes and supers (boxes to add as the colony grows) I was standing next to. Only then did I understand the name of the shop: Trees 'n Bees. Elated at this sudden stroke of genius gift-giving, I hugged Bill. The rest of the kit consisted of a smoker; a veil and cap; a pair of long, thick gloves; a hive tool; extra supers; a small book, *First Lessons in Beekeeping;* and the promise of a small wire box filled with worker bees and one queen come spring.

The bearded salesman, who reminded me of a bear, rang up our order, then showed us the observation hive on view from inside the little shop. Behind Plexiglas we could see a seething mass of bees moving along a dark-colored honeycomb. I inhaled the scent surrounding the box; it was a richly textured odor—sweet nutmeg and new wood.

I had been in love with the idea of beekeeping—danger coupled with hard work blended with sweet rewards—but figured that I could never do it in the city. My mom's friend Lowell had been a beekeeper in Idaho. I distinctly remember a trip to his farm, a land of rolling gold hills dotted with dark pine trees and white painted boxes, which my mom told me were bee houses. Lowell had wild blond hair and an unruly beard, and he had studied agriculture at Cornell before going back to the land, so he had a leg up over most of the other hapless hippies struggling to live off the land. His bees' honey came suspended in comb. The sweet golden liquid was the best thing I ever tasted. As a child, I never thought about the details. It was simple: Lowell made honey. And the idea of becoming a beekeeper myself? That seemed wildly improbable, about as attainable as becoming an astronaut.

Until Bill started to tell me about hobbyist beekeepers.

One of whom was Sylvia Plath. The daughter of a beekeeper, she and husband Ted Hughes kept bees during the happy years. Bill showed me her bee poems, and they took my breath away. "The Arrival of the Bee Box," for example:

> *I ordered this, clean wood box*
> *Square as a chair and almost too heavy to lift.*
> *I would say it was the coffin of a midget*
> *Or a square baby*
> *Were there not such a din in it. . . .*
> *I lay my ear to furious Latin.*
> *I am not a Caesar.*

I have simply ordered a box of maniacs.
They can be sent back.
They can die, I need feed them nothing, I am the owner. . . .

And Bill pointed out that there were beekeepers in cities like Paris and New York City. The forage, I read that winter in anticipation of receiving my bees, was better in cities because of city gardeners who keep plants that bloom year-round.

That spring, I returned for my bee package—a shoebox-sized cage with wire mesh sides so the approximately five thousand bees inside can breathe—and the bearlike man took me and a few other customers out to a field to demonstrate how to "install the hive." I stood in the lush green grass, terrified behind my brand-new veil. The bee guy, wearing shorts, gave a rambling discourse on beekeeping while he poured the new bees into the hive. The other newbies and I stood very far back. But as he got more and more animated about beekeeping, about the order of the hive—workers, drones, and queens—we all crept closer and closer to him. Bees landed on our shoulders and veils and then flitted off. As the details of the mysterious honeybee filled the empty beekeeping section of my brain, I felt lucky and giddy, as if someone had shown me a secret door.

The Trees 'n Bees guy did make it look easy. Then I was sent home to do the same with an increasingly angry-sounding hairy herd. I experienced a glimmer of what it must feel like returning from the hospital with a baby.

As I pulled the Dodge Dart away from the forest of Christmas trees I wondered how the bearlike man could trust me with the care of this thing. What if I dropped the box? What if they grow up and decide to swarm, to abandon me? And I thought about getting stung. A lot. Mostly because I was actually choosing to get stung. It felt a bit transgressive.

It certainly seemed so to my next-door neighbor in Seattle.

"You should move to the country," Tudy said when she saw the buzzing shoebox. She was out on her lawn, trimming the grass with a pair of scissors so that it was perfectly even. Next to her painstakingly manicured yard was our parking-strip garden, a raised bed with tall stalks of fava beans and a chaotic jumble of lettuces and Swiss chard. She hated us.

Seattle's city code allowed for beekeeping if the distance from the hive to neighboring structures was at least fifty feet. By hosting the hive on our upstairs deck, we were complying with the code. And so I ignored our neighbor and marched upstairs, clutching the bee package as if I knew what I was doing.

Then I put on as much clothing as possible. Triple shirts, a mechanic's jumpsuit, several pairs of socks hiked up and tucked into the jumpsuit, the heavy-duty-fabric beekeeping gloves (regretting I hadn't traded up for the more expensive leather ones), and finally, my veil. Swaddled as I was, I could barely put my arms down. I grabbed the gleaming hive tool—it still had the price tag on it—and installed my hive.

The sun was going down on a rare cloudless April day. Bill watched from a safe distance. As instructed, we positioned the hive to face east so it would get early morning sun. Installing later in the day avoids confusing the bees, who should spend at least a night in their hive before venturing out. I pried the lid off the bee package and tilted the opening toward the virgin hive body, with its orderly rows of frames that the bees would fill with honey.

The bees came out like a liquid, spilling into the box without incident. The Trees 'n Bees guy had showed us how to tap the package like a ketchup bottle to get out the last of the stragglers. From fear and sheer clothing volume, I had a slick of sweat dripping down my back. My terror was unfounded: The bees were entirely docile.

I fished the queen chamber out of the almost empty wire box. A few bees, her attendants, clung to the outside of the little box within the box. At the bottom was a plug of candy. The idea is that the workers will eventually chew the candy and release the queen. But I wanted to see her. So with the end of my hive tool, I somehow popped the candy inward, and she emerged. Her ass was enormous; she looked like some kind of exotic beetle. As I held the little box across the top of the beehive she strutted into her new home. Was it just me, or did she actually have an air of royalty? Then she was gone, down into her chambers, where she would lay all the eggs to keep the hive going.

I received one sting on my ring finger.

We had two years of productive beekeeping in Seattle. Bill and I worked the hives together, giving the bees sugar water to get them through the winter, adding new supers during the honey flow in summer. We harvested by stealing a few frames at a time and letting the honey drizzle out into a large pan.

Over those years, Bill and I both grew a little fatter. When I first met him, Bill was a skinny poet. Over the Seattle years he went to mechanic school at a local community college, and all that wrench-turning (and my cooking) bulked him up. I gained a few pounds, too. Maybe it was all that honey harvesting, but I think it was just being in love.

When we decided to move to Oakland, we entertained for a brief instant the idea of bringing the bees with us in our van. Using our good judgment for once, we left them with our roommates at the Hen House.

It wasn't until that second spring in GhostTown, when I started to feel like the lot might be mine forever, that we got another hive of bees. I had called our roommates in Seattle, and they had told me the news: My bees had finally died. Because beekeeping equipment is expensive, I hired some movers to bring down the empty bee boxes from Seattle. Then I ordered another package of bees like the ones I got from Trees 'n Bees. Instead of picking them up at a local bee store, I got them through the mail.

I received a desperate phone call from the post office when they arrived.

"Ms. Carpenter?" the lady on the other end of the phone panted.

"Yes, speaking."

"We've got a—what do you call it?—a box of bees, and they're freaking everyone out." It was the Oakland postmistress calling from the Shattuck Avenue office. "Can you come collect them right now—before we close?" she begged.

"OK, I'll be there in fifteen minutes," I said.

"They're outside. They've attracted all kinds of bees."

When I pulled up on my bike, a few stray bees were bobbing around the post office, undoubtedly attracted to the powerful pheromones the queen emitted from the mesh box. It was April in Northern California, arguably the best month in terms of weather. I filled out some paperwork regarding my identity, then went around to the back and picked up the humming box.

"Now, I wouldn't mind some honey next time you come by," the post-mistress yelled from a safe distance. Yup, that's most people—scared of bees but drawn to honey.

The package fit perfectly in the basket mounted on the front of my bike, and I proceeded to ride down Telegraph Avenue, laughing out loud at the bees who tried to follow us amid the traffic. At stoplights I looked down at the mesh box, the bees churning around, and told them to get ready for GhostTown.

Back at home, I placed the package of bees on the deck, then got to work setting up the new hive. (The garden would have been a better location, but I worried about the reaction of the owner of the lot, Jack Chan, to a box of stinging insects.) I placed the stand and bottom board on a table, then added the bottom box with its ten empty frames. I positioned the hive facing east, toward Highway 980, the BART trains, and, farther out, the Oakland hills. Then, wearing just a T-shirt and shorts, I casually shook the bees into their new home, fished out the queen, and placed the lid on top of the hive.

The next morning, I monitored their progress from my desk in the living room. They were circling, figuring out the new coordinates of home. *The ABC and XYZ of Bee Culture* calls these "play flights"; they establish where home is in terms of the orientation of the sun and sky. As they returned for the evening the bees were like flecks of gold, backlit by the sun. One night a few days later, I went out to the box and heard strange noises—blips and buzzes, whines and hums. When I touched it, the hive was warm, like a body.

✽

CHAPTER FOUR

A year after getting our Oakland hive, Bill and I sat in the living room and paged through the Mann Lake catalog. "The electric one is $799!" Bill exclaimed. We were hunting for a stainless-steel, hand-cranked honey extractor. In Seattle we had used a bucket and gravity to extract the honey, but the ants in California made this impossible. Normally foes of catalog shopping, we made exceptions when it came to gardening and beekeeping supplies.

"We don't need an electric one—this hand-cranked looks good," I said, reaching over his shoulder to point to the most inexpensive model.

"Looks cheap," Bill said, his big hands curled around the catalog. He pointed out that the handle on top might break off.

I moved my foot from the couch to the floor. I heard the crunch before I felt the sharp pinch on the soft pad of my big toe. My favorite description of a bee sting comes from Maurice Maeterlinck, who writes of a sting in his hilariously dramatic *Life of the Bee* as a "kind of destroying dryness, a flame of the desert rushing over the wounded limb, as though these daughters of the sun had distilled a dazzling poison from their father's angry rays." Yes, my bee sting hurt. No one was safe in our living room. Bill had gotten stung on the head. A visitor had had a bee fly down her dress and sting her bottom.

I yelped, and Bill, knowing immediately what had happened, shook his head. I curled my foot up onto his lap, and he found the stinger. The flattened bee lay on the ground where I had stepped on her. He scratched off the stinger and showed it to me. It was black and pointed, with a clear sac connected to it. It was still pulsating.

Every night in the summer, five to ten bees would sneak into our house, hell-bent on reaching the blazing lights on the fixture mounted on our living room ceiling. They came through a crack in the door to the deck. Once inside, they flew straight into the light fixture (which they might have mistaken for a cheaply made, four-headed sun). Then, stunned by the impact, they plunged back to earth. On the floor, they would crawl around in circles until they regained the strength to try again. Like the poor Icarus I had just stepped on. These nighttime escapades were an argument against keeping bees on the deck. But during the day, I liked watching them come and go as I worked at my desk.

I held a piece of ice on the sting. Bill, barely taking notice of the swelling on my toe, circled a midpriced stainless-steel hand-cranked extractor in the middle of the catalog. "Let's get that one," he said.

We were moving up in the urban farming world. The honey extractor would soon be ours. There's a saying: No gear, no hobby. The longer we lived in Oakland, the more garden-related gear we seemed to accumulate.

Later, with Bill's help, I hobbled over to Lana's for her weekly variety show. Lana's warehouse was dim and cold with a warm center. The exterior of the building was lined with corrugated metal painted a dusty yellow. You walked past the chain-link gate, through a thick metal front door, then squinted or felt your way along a dark concrete corridor that smelled like vermin. Lana, a vegetarian, loved all animals and refused to put out rat traps. Turning the brass knob on the second wooden door to the right, you fell into Lana's rowdy Wednesday-night speakeasy. Music blasted from the room, which was a riot of color, with half-finished art installations leaning against the walls. A collection of characters—old guys who grew up in Oakland in the 1950s, sculptors who worked for Pixar, buskers, and hustlers—sat at the long wooden bar.

A woman known as Bunny sat on the white leather couch. She was explaining to a sharply dressed man wearing a 1940s suit about her female wrestling troupe, the FFF. "It stands for whatever," she said, "fierce, fabulous fighters, maybe." Maya, Lana's guinea pig, sat in Bunny's lap. The guinea pig had free rein in the warehouse. Tiny brown pellets rolled on the white leather.

"We'll bust through that," Bunny said, pointing at a six-foot-tall painting on paper of the silhouette of a woman warrior with "FFF" written across her chest.

"And those are her . . ." The retro man seemed embarrassed and shifted in his seat. Maya turds rolled around.

"Yeah, her lips," Bunny said, referring to the silhouette's enormous labia.

Lana popped popcorn and poured $2 glasses of wine from behind the bar.

In the corner by the fake fireplace, Craig and Phil discussed refurbishing real wood car dashboards.

Taurean, a recent transplant from the South, explained the word "buggy" to a northerner. "You know—a shopping cart!" he exclaimed.

Bill and I sat at the bar and drank from a bottle of tequila we had brought in, with limes from our tree in the lot. We put a few dollars in the tip jar for popcorn. Everyone was smoking.

Around 11 p.m., Lana picked up her accordion. She slowly played the opening bars to a song she had learned in Spain, then she played a little faster. Cigarettes were snubbed out, drinks dashed. She played even faster as she ascended some dangerous homemade stairs, her platform boots disappearing completely. We all trailed her like rats following the Pied Piper. Maya stayed on the couch.

Upstairs Lana had set up small café tables with flickering candles and a stage with a proper curtain and a backdrop painted black. Whoever wanted to perform on this stage did.

Taurean, the southerner, who also happened to be a gay teenager, took the stage first and did some impersonations of Oakland prostitutes.

Bill and I performed a Ween-like song—he on the guitar, me singing.

The crowd hooted and yelled. They were always drunk and generous.

Another performer did a puppet show with her Barbie collection. A few country-looking people played guitar and sang folk songs. One woman wore a pair of cowboy boots. Another had her hair in braids and wore a straw hat; she had the bluest eyes I had ever seen.

At the intermission, after Bill had already stumbled home, I approached the country people.

"So . . . ," I started, using my best drawl, "where're you all from?"

"West Oakland," the woman in the cowboy hat said.

"Urban cowboys?" I said, and laughed. The country folk lived only about a mile away, on the other side of downtown Oakland.

"Urban farmers," they said, looking at one another and nodding their heads.

I hadn't heard that term before. But like I said, in California, people reinvent themselves.

"What do you mean, like milk cows and pigs?" I asked.

"No. But gardens and chickens and bees. Ducks."

I dragged them over to the lot.

"Well, here it is," I said. While they surveyed the garden beds, I noticed that the spinach looked as if it had leaf spot and that weeds had suddenly sprouted among the vegetables. I pointed up to the beehive on the deck, then showed off the crude duck pen I had made for the waterfowl.

"The chicks are still in the brooder," I said, using the word for the first time in a party setting. What I had discovered—at my various jobs, at dinner parties—was that most people didn't want to hear about my adventures in killing animals. "How can you do it?" they would ask, no doubt thinking of their pet cat or parakeet. But these people, these urban farmers, wouldn't think I was crazy.

In fact, they seemed wildly unfazed by my raising meat animals. They were doing the same thing. "We've got Muscovy ducks," said Willow, the woman with the strange blue eyes. "They're delicious!"

Willow had bought a vacant lot in 1999 and started a garden there. Eventually she founded a nonprofit called City Slicker Farm with the goal of providing healthy food at low cost to people in the neighborhood. The nonprofit sold vegetables (on a sliding scale) at a farm stand at 16th and Center streets. Willow kept using the term "food security," which idiotically made me think of chickens behind bars.

As she assessed the health of my tomatoes I told Willow about my plan to raise a turkey and eat it for Thanksgiving. She seemed impressed. "Now, that I haven't done," she said. I beamed.

"Come check out the farm stand on Saturdays," Willow said before the urban-farming entourage left.

I went back to the speakeasy and stayed late, ecstatic to have found my people. I had never met anyone like Willow in Seattle. She had long hair and wore boots, but I wouldn't call her a hippie. She got shit done, it was obvious. Like me, she was the offspring of hippies. We weren't going to make the same mistakes our parents made, I thought, taking another shot of tequila. "Viva la granjas urbanos!" I yelled to no one in particular. Wait, was it *granjas*? Something like that. Lana clinked my glass. I was an urban farmer, too.

At 3 a.m., I heard the sounds of the monks next door. Up for morning prayers, they were making clanging noises and softly chanting. Craig and Phil finally pried themselves off their barstools and began muttering about driving home. I stumbled back across the street to our house and found Bill on the bathroom floor, fast asleep.

"Bill, Bill. Get in bed," I said.

"I'm resting here," he argued. "Comfortable." His head was propped up with a folded towel. It did look kind of comfortable.

"Did you have fun?" he asked.

"Yes, I met some urban farmers."

"Those people who sang?"

"Yes, they have a garden—well, a farm like ours—on Center Street." I heard a train off in West Oakland letting out a whistle.

Bill sat up, and I helped him to his feet. We shambled to our bed, passing the glowing brooder in the living room.

I had recently expanded the brooder. The waterfowl had grown too big and were too messy to keep inside, so I had put them in a pen in the lot. But as the chicks got bigger, instead of putting them outside, where I feared they would catch a chill or get beaten up by the big chickens, I cut out more cardboard and taped on additions, until their pen took up one entire room in the house. The room filled with poultry had seemed crazy only a few hours before, but now that I had met my people—fellow urban farmers—I suddenly had a name for this thing I had been doing but couldn't quite explain. The chicks and turkey poults were still up, and zoomed around their newly expanded digs.

"How long are they staying?" Bill whispered, as if the birds were difficult houseguests. The meat-bird experiment, unlike the garden and the bees, was my exclusive domain. Bill agreed that he'd eat the birds, but the raising and the killing were up to me. I just shrugged.

At 7:30 a.m., I heard three things from my disheveled bed. One was the peeping of the chicks, who had taken to squabbling each morning the minute the sun came up. Another was the Nguyen family's morning prayers: They listened to a soothing recording of drums and chants as they burned incense. The third noise was Lana yelling in the street. I squinted at the clock and cursed tequila and Lana's speakeasy. I tucked the covers around Bill, who snored. He could sleep through violent earthquakes, the bastard.

I fed the chicks and adjusted their brooder lights. That settled them down. Then I peered out the window.

Lana was outside, causing a commotion. She looked like a comic-book character, lifting a television and heaving it at Bobby. They had never gotten along, and now that Bobby was squatting in a car just outside Lana's warehouse door, the tension had increased. Their arguments usually revolved around stuff, specifically the spoils from Bobby's demonic collecting. It was true that the end of the street was starting to resemble an open-air flea market. Bobby had mounted a corkboard sign near his home on which his friends and associates could post messages. He was a social fellow and enjoyed company. Many of his friends brought "gifts," one of which was the television currently being flung through the air.

I wandered outside to mediate.

The two were looking at a small square of earth on the curb where a tree had probably stood in the days before Oakland had gone to hell. These days the spot was home to a cheeseweed mallow, *Malva parviflora*. Though she hadn't planted it, Lana had an inordinate love for this weed. It had little pink flowers. It also had an invasive attitude and a pernicious root system that I, as a gardener, could never love.

Bobby had dug it up and planted a cactus.

God knows where it came from, but the cactus was spiny and columnar and freshly planted.

"Morning, darling," Bobby said to me.

"Hi," I mumbled, wanting to seem unbiased.

"I'm just making some improvements," Bobby explained, then pointed to the cactus.

"There was something there already!" Lana yelled. The wilting *Malva* lay over by the abandoned playfield. Instead of ripping up the cactus—Lana couldn't hurt another living thing—she had smashed an electrical appliance. The shards of thick glass and wood littered the street.

It was a turf war between two 28th Street impresarios.

Lana had been up all night. I knew because I had left her house only a few hours before, and she looked just as she had when I waved goodbye. Her hair still looked wonderful, and her eyes were painted with theatrical curlicues of black eyeliner.

"Some people don't listen," Lana said, glaring at Bobby. He sometimes posted life lessons on his message corkboard. One of his favorites was "Learn to Listen."

"I'm just trying to help," Bobby said.

"I don't want your help!" Lana yelled. Her dog, Oscar, wandered out from the warehouse and gave a loud bark. He liked Bobby, despite Lana's hatred, because Bobby fed him snacks of old bread and bones. This further infuriated Lana.

"Some people need to learn how to relax," Bobby said, drawing out the last word.

Bobby had recently started an auto-repair/chop shop—a place to strip cars—at the end of 28th Street. It all started when a neighbor said her car was dead. Bobby opened up her hood, then rummaged around in the back of his car, emerging with a car battery. "I went to Berkeley," he said to the woman. "Studied biology." He flipped the battery upside down so the terminals touched. Battery acid flew. Her car started. Word spread: Bobby can fix cars.

When it came to my loyalties, I sat on the fence. On one hand, I was like Bobby in that I was running what some might consider an unsanitary operation (horse manure, chicken shit) on squatted land, and Bobby guarded that land for me. On the other, I was like Lana in that I enjoyed an environment free of antifreeze spilling into the gutters and an endless collection of Ghost-Town rubbish. (Bobby once found a giant metal shed and placed it in the lot. When I protested—what would Jack Chan say if he saw that?—Bobby took it away, grumbling.)

But now, in the battle between a cactus and a weed, I knew I couldn't choose. This weed, *Malva parviflora,* was truly awful, and a dime a dozen. As for the cactus, it did require almost no water, but having a spiny plant spiking out next to a sidewalk seemed sadistic. I couldn't referee this one, and so I shrugged and said, rather lamely, that I wished we could all get along.

A few weeks after my tequila-enhanced discovery that I was indeed an urban farmer, the chicks started to develop real feathers. Worried that they weren't getting enough vitamin D—an important nutrient for healthy feathers—I took the birds on a field trip to the garden.

I gathered the birds into a cardboard box and then upended it into a garden bed. At first they seemed stunned by the out-of-doors—the shining sun, the blue sky, the dirt at their feet. When a pigeon flew by, without a sound they took cover under the crinkled leaves of a Swiss chard plant. The turkey poults ducked their heads underneath the chicks' legs, trying to hide—never mind that they were twice the size of the chicks by then. After a few minutes, the instinct to roam trumped fear. The chicks fanned out, scratched at the dirt, and ate rocks. Chickens and turkeys don't have stomachs. Instead they digest their food with a powerful muscle in their gut called the gizzard. The birds peck up rocks, which travel into the gizzard. Once there, the pebbles are ground against one another by the gizzard's contractions, which breaks down the grains and greens and bugs.

The birds were beautiful there in the garden. The chickens ranged from dark-reddish to black to yellow-gold. The turkey poults were starting to get some color, too. One had dark black feathers; the other two were mostly white with streaks of black.

As the birds enjoyed the early summer sunshine I did some weeding and checked on my vegetable seedlings.

Not only was I growing meat birds; I had expanded to include some varieties of heirloom fruits and vegetables. I chose heirloom varieties because they are often best suited for a small home garden, because their seeds can be saved and used the following year, and frankly, because I love their names: Amish Paste tomatoes, Golden Bachau peas, Speckles lettuce, Cream of Saskatchewan watermelons. All seemed to be doing fine except for the watermelons.

In early June I had made careful mounds of composted horse manure and dusted the tops with worm casings harvested from the worm bin. Then I sank the watermelon seeds into each mound. I watered them very well, resubmerging two escapees that floated up to the surface.

I anticipated sweet melons that I would eat straight from the garden, juice dripping down my shirt. Since I had moved to California from Seattle, watermelons had held a special cachet for me. They are native to Africa and require heat to develop. The Pacific Northwest just didn't have it. In Oakland, I had so far mastered hot-weather-loving plants like tomatoes, hot peppers, and corn. This was my chance to do melons.

For a week, I had gone out every day to water and stare at the dirty black piles, to see if anything was emerging. On day three, I thought I saw some green, but it was just a piece of trapped grass. On day five, I sprinkled the mounds with fish kelp fertilizer and suppressed the urge to dig in to see exactly what was going on down there. Eight days had now passed, and I cursed the seed company, birds, bad horse manure, ants, and any other suspects who could be blamed for preventing the watermelons' germination.

I pawed through the dirt to find some potato bugs for the ducks. In four weeks, the ducks and one of the goslings had made it to full feather. (The other gosling died quietly in the brooder one night for no apparent reason.) Because they had grown so quickly and had such wet, fly-attracting turds, I had moved them outdoors much sooner than the chicks, into a pen I built out of chicken wire and milk crates. I had even made them a small pond—a washtub sunk into the dirt and filled with water.

A few feet from the pen was a chain-link fence, and behind that was a duplex—a cobbled-together adobe affair—where many people lived. My favorite residents there were a woman named Neruda and her nine-year-old daughter, Sophia. When I first put the ducks out into the yard, Sophia watched from behind the fence, too shy to say anything. After a few days, she spent more and more time lurking and watching the birds' antics. One day I invited Neruda and her daughter over.

The ducks—Pekins, a popular breed of meat duck from China—were almost fully grown by then. They had glossy white feathers and sturdy orange bills. The surviving gosling had turned into a stately gray goose. We sat in the sun and watched the ducks play. Sophia picked up one of the lily-white birds and gently set him in the "pond." He quacked happily and bobbed his head in and out of the water.

"Is he like a pet dog?" she asked me suddenly.

I glanced at her mom. Cornered.

"Not really," I said, feeling like a monster. While Sophia had been playing with the ducks, I had been thinking about duck confit and Christmas goose.

In fact, I had been studiously avoiding the thought of killing, focusing instead on the first few sections of *Storey's Guide to Raising Ducks,* which told me how to install a pond and what to feed the growing flock. I hadn't gotten to the butchering section of the guide yet. Nor had I yet gone to Willow's farm to find out how she killed her ducks. Next to *Storey's Guide* on my nightstand was Elizabeth David's *French Provincial Cooking.* I read with keen interest her many good ideas for cooking duck livers and making *canard rôti au four.*

But how do you tell a child you're going to cut off this adorable duck's head, pluck its white feathers, and roast it in an oven, letting its fat naturally baste the meat? Children, I've found, don't care much about haute cuisine. So I looked into Sophia's innocent chocolate-colored eyes and mumbled something about eggs and breeding. Yes, I straight-up lied.

After Sophia went home, I sat in the lot and looked at the ducks. There was an unmistakable gap in my knowledge of these creatures, right there in between the raising and the cooking. I knew how to raise them, and I knew how to cook them. How to get from a living duck to a duck ready to go in the oven—that was the trick. Not only did I not have the physical, practical know-how; I didn't know how to prepare myself mentally, either. I suspected that there wasn't a single book that could fill the gap between *Storey's Guide* and Elizabeth David.

While the chickens and turkeys foraged in the garden bed, I leaned into the pen and offered the ducks the potato bugs. They softly prodded my open hand and snarfed them up one by one. Then they looked at me for more.

I went back to the doomed watermelon area to hunt. As the ducks quacked and chortled at me and monitored my movements with great interest, I scratched the soil around the edge of the bed and found a roly-poly paradise, with bugs everywhere, in every size—from tiny ones the size of a speck of dirt to big ones the size of a cockroach.

A faint glimmer of green made me halt my potato-bug harvest. Hidden under the dirt, seedlings had been growing. They looked like most melon sprouts do: rounded, kind of veiny. My watermelons had finally germinated.

I crawled closer to the seedlings and inspected them in worshipful silence. Two were only half unfurled (the seed coat still hung on one of the leaves), another was a bit off to the side of the mound, a fourth was small and runty. One bruiser sat smack-dab in the middle of the pile. I swear I could hear it growing. Seeing the watermelon seedlings felt like finding money in the street: even though I had done the hard work to set the plant in motion, it still seemed like a miracle.

When seeds germinate, an amazing thing happens. A seed is a ripened ovule, like a hen's egg: it contains an embryo and a stored food supply. I watered the seeds every day because of a process called imbibition. When a seed soaks up water, its cells swell and mitochondria (the power stations of cells) become rehydrated and start to work. A cascade of proteins is made, the food-storage reservoirs are tapped, and slowly the cell wall softens. As cell division begins—set in motion by the rehydration process—a radicle bursts out of the seed coat and becomes the root of the plant. All this had finally happened to my Cream of Saskatchewan watermelon seeds.

After an hour outside, I shooed the chicks and poults back into the cardboard box and took them upstairs. I could see the veiny leaves of the watermelon from our living room window. The process was in motion—all I had to do was give the plants regular water, perhaps side-dress with some compost, and hope my bees were up for pollinating a watermelon flower. I would soon be the proud eater of a homegrown, rare-breed watermelon. Nature had succeeded, despite the odds, again. Even in a plot next to the highway, germination is possible.

The next morning, I stood in my lot and yelped. I threw the hose off to the side—I had been watering—and examined the crime scene. Half my watermelon seedlings were now stubs. My eyes trailed a jellyfish-like slime that ended at what was left of the baby melon plants. Chewed by slugs.

Searching for the culprits in the sunlight, I dug around the moist areas where soil came in contact with wood, a slug's favorite hiding spot. I found a few small ones. They looked—and felt—like pieces of gray snot. Now that I had them, I weighed my options. Some people suggest tossing slugs; that is, you just hurl them as far away from your garden as possible. I had my doubts that the greedy mollusks deserved a second chance, however slim, to slowly creep back from the street, dodging cars and boots, and snack once again on my delicate watermelon seedlings. Other people suggest drowning them in beer moats crafted out of tunafish cans. The slugs would fall into the moat and die a drunken, Janis Joplin–esque death. This seemed suspiciously close to buying the slugs a beer, which was more generous than I felt. So I dispatched them between my thumb and index finger. I offered their corpses to the ducks and the goose, but they didn't seem interested.

I knew there were more slugs—bigger slugs, the mothers and fathers of these babies I had just murdered—and I knew when to catch them. Later in the day, as the sun set, I drank a strong cup of tea and strapped on my headlamp. Prepared for hand-to-hand combat, I went slug hunting.

I found them, lit up by the strange blue light of my energy-saving headlamp. Lolling around in the dirt, they approached the supple green shoots of my remaining melon plants with their little horns (which are, in fact, eyes) up. Some people think the horn-eyes are endearing, but if the slugs had their way, I would have exactly zero watermelons. These dirt-inhabiting scum balls are the clear-cutters of the mollusk world. They will leave nothing in their wake. If they exercised restraint, they could have a food source for months instead of hours. But this isn't how slugs operate. And so I committed slug mass murder in order to save a fruit.

I ripped them in two; then, just to be sure, I squished them between boards. When it was done, I tossed the grotesque, loogie-like pile of dead slugs into the garbage can.

As I scrubbed my murderous hands I wondered if Lana would refuse to eat one of my watermelons if she knew what I had done. No one is really pure, except maybe the Jains, that sect in India whose members sweep a peacock feather on the ground in front of them as they walk in order to prevent injury to, say, an ant. They do drink water, I'm told, and I wonder how they come to terms with eating all the organisms that live in it.

It took me a full five minutes of scrubbing to remove all the slug slime from my hands. Like a low-stakes Lady Macbeth, I couldn't shake the sensation that they were still soiled. But I wasn't conflicted: I felt great. I killed so that others might live. Death is all around us, even in an innocent watermelon. You just have to know where to look.

CHAPTER FIVE

A visitor used the word "unhygienic" to describe the almost full-grown chickens and turkeys living in our house. She had a point. Our record player was coated with a golden dust I had never seen before. Bill complained about the noise they made—even he, deep sleeper extraordinaire, couldn't sleep through their crack-of-dawn racket anymore. When I read something about getting chest infections from living in close quarters with chickens, I finally moved them outside.

I was reluctant for a reason. We already had chickens. Since they are territorial and had an established pecking order, it was going to be a brutal smack-down for the little ones. So I slowly introduced the turkey poults and the new chickens to the big chickens. First I kept the new ones in a large wire pen. The big chickens thought this was some kind of gladiator event and lined up breast to breast to peek through the wire at the newbies and get in a sucker peck whenever possible. After a few days of this, they all knew one another, and I unleashed my chickens and turkeys into the cruel world of urban chickendom.

They fared well. A few of them got their asses kicked. One of the big chickens, a beautiful, normally docile Buff Orpington, body-slammed one of the poults and mercilessly pecked its head until it yelled uncle in turkey. That's how chickens do it. They establish dominance, an order, that every bird agrees upon, and then they get back to what they do best: pooping and eating.

Back in Seattle, our first chicken was an Americauna named Agnes. She was a lesbian chicken who crowed like a rooster but also laid eggs with bluish shells. At the time—the late 1990s—I was working for a company that published a book called *The Encyclopedia of Country Living,* by Carla Emery, a how-to guide for would-be homesteaders. I laughed as I thumbed through the book in my cubicle in downtown Seattle. How to dig a root cellar, shoot a pig, and castrate a goat—not things that I would be doing anytime soon. But the sections on how to can vegetables, grow pumpkins, and keep chickens—these were things that even a city person with a backyard could do, I realized.

Keeping backyard chickens was more than socially acceptable then. Martha Stewart had them; PBS aired a documentary about poultry fanciers; and in Seattle, having chickens in the city was a badge of progressive moxie. Finally driven over the edge by Emery's book, I bought Agnes and three full-grown Golden Laced Wyandottes, beautiful gold- and red-feathered chickens who laid big brown eggs.

These hens provided more eggs than we knew what to do with. I started to breeze past the egg section in the supermarket. Cold, white eggs were an affront compared to our warm brown- and blue-shelled eggs, so fresh they didn't even need to be refrigerated.

After two years of egg-laying service, Agnes was killed by a dear friend's dog. We held a formal funeral, during which I distinctly remember swooning. A few weeks later the same dog dug up Agnes from her final resting spot and presented horrified backyard picnickers the ossified corpse as a gift. My beloved chickens were pets, almost human; I had never thought of them as meat birds.

What I didn't know then is that keeping chickens in Seattle had placed me squarely on the path toward urban farming. Chickens are the gateway urban-farm animal. Because of them, I would soon be learning how to kill and pluck a duck and a turkey. If smoking marijuana led to snorting cocaine, then chickens eventually led to raising meat birds.

After I released the birds into the backyard, I officially updated our chalkboard tally:

3 turkeys
3 ducks

1 goose
14 chickens
50,000 bees (they were doing really well)
74 flies (ditto)
2 monkeys

A few days later, I came outside to find Lana, her sister, and her guinea pig, Maya, in the backyard. They had come over to see the new feathered flock. Lana was in love.

"Wow, who's that?" she shouted, pointing at the black turkey as she squatted down to admire his shiny, iridescent feathers. Glad to get attention, he puffed them all out and made a huffing noise. His black tail feathers stood at attention.

"That's one of the turkeys," I answered. "He doesn't have a name," I added pointedly—real farmers don't name their meat animals. Another turkey, a small white and black female, was eating some corn. I couldn't find the other male. Now that I thought about it, I hadn't seen him all morning.

The preening black turkey glided in front of Lana. His head blushed blue. The sunlight made his feathers glow. "I don't want to hear why," she said, not looking at me.

Lana was a strict vegetarian. I understood—I had once lived a meat-free life. Starting in high school with the refusal of a steak, I had forced my sister and mom to go vegetarian with me. I happily ate cheese sandwiches through my first two years of college and dutifully made earnest, bean-heavy meals from the *Moosewood Cookbook*.

My fall from grace came in Las Vegas. There with friends over a college spring break, I looked at a Circus Circus breakfast buffet that included a ceiling-high stack of bacon and felt dizzy with desire. My years of resolve floated away, and I ate fifteen pieces in one sitting. I felt simultaneously awful and wonderful. Though the top of my mouth felt as if I had eaten a can of Crisco, all that protein gave me vivid dreams, and I had the energy of one of the Bull Ship Pirates from the hourly Treasure Island show.

The next day, strolling down the Bellagio's fake Greek pavilion and thinking about my next meat binge, I started to worry about the origins of the pig meat I'd eaten.

The PETA videos and the anti-factory farm comic books that had been my vegetarianism's inspiration weren't easily forgotten—the wheezing pigs getting slapped around by mean (and probably underpaid) workers; the live baby chicks piled on top of one another in the Dumpster; the turkeys hanging from a conveyor belt as workers slit their throats one after another, as casually as turning a page in a book. In these settings, living beings—animals who love sunshine, fresh food, and taking naps in hay—became meat machines. In meat factories, the animals weren't allowed to be truly alive, and that was wrong. Lana and I agreed on this point. However, I couldn't believe, as Lana did, that animals were like little people wearing fur coats.

Lana held the guinea pig up to the turkey. "Harold, meet Maya," she said.

"Harold and Maude," Lana's sister said, laughing.

Suddenly my turkeys had names.

There's a weed called pellitory that grows all over GhostTown. It can grow in the tiniest crack in the sidewalk and flourish. My chickens loved it. I noticed this fact when we put our first four Oakland chickens in our pellitory-choked backyard and they chewed down every shred of the weed they could find.

I then picked all the pellitory that grew in our lot, and the birds literally came running toward me, they loved the snack so much. With fourteen chickens, plus the turkeys and the waterfowl, who loved pellitory, too, I desperately needed another source. Luckily, all over our neighborhood, pellitory grew alongside houses, in lawns, and through chain-link fences in abandoned lots. It was there for the taking. But first I would have to get over my fear of walking around our neighborhood.

When we first moved to GhostTown, I wouldn't walk around at all. Our landlords lived four blocks away but insisted, for safety's sake, that we mail the rent checks. Lana had been right that the 2-8 was like Sesame Street, but beyond it, all bets were off. GhostTown was a gauntlet of crackheads, homeless guys, and prostitutes. There were drive-by shootings almost weekly. When venturing out, I either rode my bike or drove my car. I never walked down the streets.

But I had noticed a patch of pellitory growing along the abandoned brick building on the corner. And I had read in *The Encyclopedia of Country Living* that meat birds who eat greens will taste better. I became motivated. I stuck a toe in: drifted to the end of our safe street and found myself on the main drag, where the weeds grew and often bullets flew. As I pulled up weeds various bleary-eyed citizens wandered by, stared at me for a minute, and then said hello or good morning. Even people I had written off as totally fucked up—like the scabby blonde who was always spare-changing everyone—were quite friendly. I was a little ashamed that it had taken me two years to finally venture out into our neighborhood on foot.

On a bike ride the next day, after my successful harvest, I happened to notice the pellitory on 29th Street—a busy thoroughfare that attracted a lot of pot-smoking teens. There was a constant layer of debris on the street, and at night dark-windowed cars idled on all the corners. But 29th also had Durant Park, a little green spot where the pellitory grew lushly. In order to get to these weeds, I had to get over my fear of the guys who presided on that corner.

One night, drunk at the Blue Wednesday speakeasy, I explained to Lana my problem: wanting the weeds but fearful of the thugs. Lana worked at a local teen drop-in center and had lived in our neighborhood for fifteen years. She knew everyone, even the guys who intimidated me.

"They're just babies, Novella," she said to me. "Imagine growing up and everyone is scared of you. Pretty soon you use that power—you become what everyone is afraid of."

Bolstered by Lana's insight, the next day I gathered two plastic buckets and went out for my maiden voyage to 29th Street. As I rounded the corner of 28th Street I took on my best don't-fuck-with-me attitude. At least I was trying, though I'm not sure anyone carrying two clattering buckets can be entirely tough. I walked down the sidewalk. A group of ten guys slouched on the corner, blocking my way. "Babies, babies, babies," I muttered to myself.

"Excuse me," I said. "Hello."

The group parted, and a few of them said hello back.

As I walked through the crowd I looked up at the tall teenagers and smiled. Lana had been right. I harvested a shitload of weeds at the park and brought them home for the birds. On my way back, I realized that, carrying the buckets of weeds, I must have appeared just as crazy and eccentric as anyone else on the streets of GhostTown.

One hot morning in the middle of summer, I grabbed my buckets and headed to the park. As I clattered past Brother's Market the shopkeeper, Mosed, waved from behind the counter.

I turned onto 29th Street. None of the teen guys were on the corner. At Durant Park, a circle of candles burned on the sidewalk. Just the day before, the sound of gunshots had echoed through the neighborhood. Bill and I had stopped what we were doing—he was working on a car, I was sowing some lettuce seeds—and peered down MLK. Police cars came, then an ambulance. Now a T-shirt, attached to the park's gate, "Rest in Peace 1985–2005" written on it with a Sharpie, marked the death. A few teddy bears and empty Jack Daniel's bottles sat next to the candles.

I stepped past the altar and began gathering weeds. A little kid in diapers across the street watched me from behind a gate. Pellitory is soft green with red stems. Young and pliable, the stems break off easily, but the plant's strong root system ensures its survival.

I worked the area underneath a eucalyptus, pulling up handfuls of the weed. I wondered whether Lana had known the victim. He would have been five when she moved here. She probably saw him riding his bicycle around. And then, with no opportunities, he eked out a living on the corner. Maybe spent some time in jail. Who knows what happened.

Tiny burrs from the pellitory dusted my sweatshirt sleeves. My hands were slightly wet from pulling up the dewy plants. I had two bucketfuls, plenty. I waved goodbye to the baby at the gate, then turned the corner and walked back past Brother's. Two loud men paused outside the store to crack open their brown-bagged cans of beer. It was 8:30 a.m. After seeing the altar, I could understand the logic of such a decision.

I put down the buckets and went in. The store had two aisles. Gum, candy, chips, cans of beans, and plastic bags of pasta were on one shelf; the other was devoted to alcohol: jugs of Gallo wine, Wild Irish Rose, Boone's Farm. I grabbed a six-pack of Tecate—for later—and placed my purchase on the counter. Two Yemeni men sat there; behind them were batteries, phone cards, and cigarettes.

Mosed smiled and rang me up. He has dyed red hair—vivid red, not natural at all—and a goatee. His wife, in a head scarf, stood in the doorway that led to their apartment upstairs.

As I handed Mosed my money the irony of buying alcohol from a Muslim man wasn't lost on me. He nodded and passed me my change—he had stopped judging his customers years ago. The bills were worn ghetto dollars, as thin as Kleenex. I nestled the beer in one of the weed buckets.

At home, the chickens and Harold and Maude had fanned out in the backyard; they kicked up mulch, took dust baths, and fought over unearthed bugs. When they saw me, they came running. I know it's pathetic, but to be loved, even by poultry, felt great that morning. I threw the weeds into the chicken house. When the hens and turkeys enthusiastically followed their favorite treat, I shut the door behind them. Then I lumbered upstairs with my six-pack and went back to bed. I locked the front door with all three locks. And the chain.

Ten blocks from my house, I found Willow's farm and garden. An orange sign read CITY SLICKER FARM in purple. The Center Street garden, just off 16th Street, burst with vegetables and fruit. A pen of ducks and chickens straddled the back of the property. A chayote, a vining squash, covered the entire front fence. Tall columns of peas stood guard near the gate, with strawberry plants at their feet. Tomatoes had been staked and supported.

Willow had her head in a giant outdoor oven made of adobe. I hadn't seen her since the night of Lana's speakeasy, but she had been on my mind as the ideal urban farmer. As I tried on that identity, Willow was my model.

"Hey, Willow," I said, feeling a little shy.

She jumped, pulled her head out of the oven, and said hi. She gave me a hug ("Sorry, I'm a Californian, I hug everyone"), then took me on a tour of the garden.

"This soil was full of lead," Willow explained, showing me the raised beds. Her garden looked very much like a mature version of what I hoped mine would eventually resemble.

"But what about the fruit trees?" I asked, pointing at the fig and mulberry trees.

"We had the fruit sent to the lab," she said, "and the fruiting bodies don't contain lead. The leaves do, though." The leaves, which pulled the lead out of the ground, were hauled to the dump. Every year, the soil was getting cleaner. The garden, then, was a giant remediation project.

After she showed me the bees, the chicken house, and the toolshed, Willow went back to making a fire in the oven. She was going to make pizza for the neighborhood. Somehow she had gotten all the fixings—the dough, the cheese, the tomato sauce—donated for the event.

One of Willow's volunteers set up the produce stand with a colorful display of beets, chard, and carrots. A small basket of lemons and figs. A few live plants for sale. Customers stopped by to buy vegetables and were invited to eat some pizza. One man only spoke Spanish, and Willow came out to the stand to talk to him. He wanted to buy some ducks for a duck roast. They made arrangements.

I wandered through Willow's garden, admiring the construction of the chicken pen, the beehives. Outside the gates of her farm were crumbling industrial buildings. A man pushing a shopping cart onto a nearby lot paused to take a piss on one of the buildings. I couldn't help but think of Wendell Berry, the strident agrarian. Not that he would pee on a building, but what would he—all rural values and fan of sweet-smelling fields—make of this farm? Berry clearly hates cities. "No longer does human life rise from the earth like a pyramid, broadly and considerately founded upon its sources," he wrote in *The Unsettling of America*. "Now it scatters itself out in a reckless horizontal sprawl, like a disorderly city whose suburbs and pavements destroy the fields." Cities destroyed fields. The soil under my favorite bar could be growing corn. That art museum? Just a platform of concrete.

But not all of us can live in the country like Wendell Berry. Of course he knows this. In perhaps his most famous essay, "The Pleasures of Eating," Berry advises city people, "If you have a yard or even just a porch box or a pot in a sunny window, grow something to eat in it. Make a little compost of your kitchen scraps and use it for fertilizer."

Or, if you're Willow, you might do a little bit more than that: Create a farm in a city lot, sell produce on a corner, show urban kids where eggs come from. Plant in the cracks of the city.

This idea isn't a new one. Most of us have forgotten about the depression of 1893. It hit Detroit hard. Because of a bank panic, industry in the city came to a standstill. Ten percent of workers were unemployed. Food shortages threatened. A plump, balding, bearded shoemaker turned mayor came to the rescue. Hazen Pingree looked around Detroit and saw abandoned lots. Lots of them. He wondered why the unemployed should not be allowed to cultivate food on them. On his travels in Europe, Pingree had seen allotment gardens, plots of land set aside for city folk to grow vegetables and flowers. These became his inspiration. By 1896, there were Pingree Potato Patch farms all over the city. As Laura Lawson, in *City Bountiful: A Century of Community Gardening in America,* reports, in one year, "the program served 46.8 percent of families seeking public relief and the gardeners grew $30,998 worth of food." Word of the success spread, and soon New York City and Philadelphia had their own vacant-lot farming programs.

They didn't last forever. Once the depression was over, the programs ended, for the most part. But they sprang up again during the First World War, then again in the form of victory gardens during the Second World War. Flourishing, then disappearing—this has been a way of life for urban farmers in America.

Willow assigned me the job of riding my bike around and yelling,

"Pizza! Pizza! At 16th and Center!" West Oakland looked as bad as Ghost-Town, I thought as I pedaled around. Fenced-up parks, abandoned buildings, charred cars. Bored kids with nothing to do but follow this crazy lady on a bike to get some free pizza.

These kids would have few chances to experience the rural places described in Wendell Berry's books. Because of Willow, they could harvest a tomato or see a chicken lay an egg, and on a summer day they could watch the mulberry tree ripening. To be a farmer, Willow pointed out, was to share. Unlike a rural farm, a secret place where only a few lucky people may visit, an urban farm makes what seems impossible possible.

The pizzas, fresh from the wood-fired oven, had crispy crusts like those you find in Italy. Many of the toppings—basil, garlic, onions—came from the garden. It was the best pizza I had ever eaten. And when the kids on the corners followed me to this dazzling place of greenery, this place of goodness, and ate the best pizza on the planet, I fairly burst with happiness.

While the neighborhood kids swarmed around eating pizza and looking at the beehives, Willow and I discussed killing ducks. It was getting close to their time to go. I saw them as good practice before I had to do the big kill: Harold and Maude. Willow recommended using pruners. We made plans for a hands-on demonstration.

CHAPTER SIX

❋

I had tool envy. Mr. Nguyen had a thing that looked like a hoe but with a shorter handle and a deeper blade than on any other hoe I'd ever seen. We were both in the garden. He was clearing out his patch, where in years past he had grown taro, a root vegetable with enormous elephant-ear leaves; yellow chrysanthemums, whose leaves the Nguyens used for cooking; and an orange tiger lily.

I was jury-rigging the raised beds so they would be protected from the onslaught of the chickens smart enough to make their way (walking, flying, sneaking through a fence) from the backyard into the lot. Some friends who had recently moved to Portland had given me their layers, so my hen population had swelled to more than twenty and was a force to be reckoned with. The hens had recently laid into my garden with a ferocity I hadn't seen since my slug-murdering session. They kicked up my tiny, defenseless seedlings. They pecked the chard down to nubbins. They uprooted a newly transplanted tomato. It was pure chance that they didn't uproot my prized watermelon seedlings. People always say chickens in the garden keep the bugs down, but as far as I could tell, they were hell-bent on destroying everything *but* the bugs.

So I had gone on the defensive. This involved wrapping each of the raised beds with chicken wire and stapling it into place. It wasn't attractive, but it would keep the upstart marauders out.

Mr. Nguyen was busy whacking back mint and making room for more red perilla. The strange tool had a sharp edge that turned inward, so he could hack with it to dig trenches and smooth out a planting area. I went over to where he was working. He wore a pair of dress pants and a tucked-in white shirt. I asked him what the tool was called.

"What?"

"The hoe, what do you call it in your country?" I asked, and pointed down.

He said something in Vietnamese. I still hadn't mastered "good morning" or "thank you" in Vietnamese; I'm a complete moron when it comes to languages. I could tell he felt sorry for me. He smiled and said, "Hoe."

He was a natural urban farmer. Before Bill and I cleared out the lot and planted it, Mr. Nguyen had tended an herb garden in the backyard, but it never got much sun. We persuaded him to move it out to the lot.

Strolling down MLK on my weed-gathering missions, I had started to notice several places where other Vietnamese gardeners had reclaimed a patch of land in their front yard or along the side of the house next to the driveway. In one little corner of a yard a few blocks from ours, red-leaf mustard greens grew alongside cilantro, *rau ram* (Vietnamese cilantro), and lemongrass. Maybe it wasn't enough food to feed a family, but it was a taste of home.

In his delightful memoir *The Unprejudiced Palate,* Italian immigrant Angelo Pellegrini describes the newly arrived Americans of the last century: "He subsidizes his fluctuating income by wringing from his environment all that it will yield. . . . Regardless of his means, he will garden his plot of ground because he knows the vital difference between cold storage or tinned peas and those plucked from the vine an hour before they are eaten. Furthermore, challenging the soil for its produce is in his bones; the pleasure of eating what he raises is inseparably fused with the pleasure of raising what he eats." So it went for most immigrants to America: Pellegrini grew his cardoons and basil in the 1950s; the Vietnamese and El Salvadorans of this century sow cilantro and lemongrass.

As I feared, a few of the new chickens wandered into the lot. They scratched about, hit their heads repeatedly on the chicken-wire fences, and then gave up. Mostly they ignored me. Harold and Maude rounded the corner and came into the lot, chirping and barking. Harold was getting mature and had developed a major wattle. It looked as if melted red plastic had been poured over his head and solidified midpour. His snood, a fleshy piece of skin, now hung over his beak. When they saw me, they rushed over and pecked at my fingers until I had to hide my hands in my pockets. Then the birds pecked at my pant leg.

The turkeys weren't growing up to resemble the white turkeys that most American farmers raise, and they definitely looked different from my mom's turkey, Tommy. Harold was a deep black, with some white on his tail; Maude had alternating white and black feathers, like an exquisite houndstooth jacket. McMurray Hatchery, bless them, had sent me heritage turkeys.

I discovered this while working a booth at a book festival in San Francisco. I was browsing through some of our food porn: Slow Food International's *A World of Presidia*. The book featured hundreds of endangered heirloom plants, animals, and food products with close-up photos and centerfolds. Delicious vittles like Hungarian Mangalica sausage, made from a curly-haired pig; Tibetan Plateau Yak Cheese; and a Chilean Calbuco Black-Bordered Oyster.

I turned to page 90 and was delighted to see Harold. He was, according to the book, a Heritage Standard Bronze. Maude was apparently a Royal Palm. The book listed other heritage turkey breeds, like the Bourbon Red, the Narragansett, and the Jersey Buff. These breeds can be traced back to wild turkeys taken from North America, sent to Europe, domesticated into breed categories, and then sent back to the States in the late 1700s.

These heritage breeds aren't eaten much anymore. Slow Food blamed my turkeys' distant cousin twice removed, the Standard White. Turkey breeders in the 1950s wanted a standardized bird, one that grew quickly and finished with a uniform size that would mesh perfectly with new mechanical pluckers that had been developed. With careful breeding of heritage stock, they arrived at the Standard White. Over the years, the breed has been further engineered to do well indoors, and the breasts have plumped up enormously. On a strict feeding regimen, a Standard White takes just two months before he's ready to eat. He's a meat-growing machine on two legs.

My heritage turkeys, on the other hand, were growing slowly—it would take six months for them to develop fully. The difference in taste, according to the Slow Food book, makes it worth the wait. Firm, extraordinarily dark meat. More delicious breasts and thighs. They might be happier, too: The Slow Food book reported that heritage turkeys, unlike Standard Whites, can indeed mate naturally.

My turkeys were heritage as hell, I told myself as I slobbered over the book. And the fact that they could have sex was somehow wonderful news. I was going to have the most amazing Thanksgiving feast of all time. Only three more months to go.

Mr. Nguyen's wife, Lee, pulled up in their battered silver car and tooted the horn, then began to unload bright pink plastic bags from Chinatown. Mr. Nguyen finished his hoeing and unloaded crates of oysters from the trunk. The night before, the Nguyens' son had erected a giant white tent in the lot, and he was now setting up tables.

By noon, the flimsy plastic card tables were threatening to collapse under the bounty of rice-noodle salad, prawns, steamed rice, sliced cucumbers and tomatoes, and, of course, cold Heinekens. Bill and I were beckoned down from our apartment. It was the first-year birthday party for the Nguyens' grandson Andrew.

Giddily, Mr. Nguyen led me to three smoking barbecues. He pulled a giant oyster from the grill with a hot pad and pried off the top shell. While we waited for the grayish piece of protein to cool down, Mr. Nguyen demonstrated in pantomime that I would first dip it into a bowl of pepper and salt, then squeeze a bit of lime over the whole thing. It was an epiphany. The rough grit of the pepper, the sweet oyster, the sour lime—perfection. I looked over at Bill, who was wolfing down oysters with Mr. Nguyen's son, Danny, and drinking a Heineken. I noticed that none of the Vietnamese women drank beer. The birthday boy slowly shoved rice and birthday cake into his mouth.

Once Mr. Nguyen saw how much I enjoyed the oyster delicacy, he deemed me ready for the grand treat. He handed me a largish egg, still hot from the grill, and a spoon. Sensing my confusion (hard-boiled eggs?), he demonstrated that I should tap off the top of the egg. I did so, and a yellow-ish fluid came out, revealing a duck embryo floating in the yolky orb. While Mr. Nguyen watched and encouraged, I scooped out some embryo, which somehow had feathers, and gave it a taste. It was like a salty Jell-O banana pudding topped with bonito flakes.

I faked delight, thanked Mr. Nguyen, and wandered off to deposit the thing under a cabbage leaf.

"Novella, what are you doing?" Mr. Nguyen's ten-year-old granddaughter, Tammy, said as she caught me burying the culinary monstrosity.

She laughed when she saw the uneaten egg. "Don't tell your grandpa," I begged.

"They're pretty gross," she said, like a teenager, then flitted away.

Other people at the party followed my lead, except their embryos were eaten. By the end of the day, the garden was heaped with empty duck eggs and oyster shells, some of which I later used as impromptu digging tools.

After the party was over, I herded the turkeys back to their roost.

"What's that?" our neighbor the Hillbilly asked. We called him the Hill-billy (behind his back, of course) because he regularly "borrowed" packs of cigarettes from Lana, wore a camo/American flag baseball cap, and worked the night shift as a security guard at Wal-Mart. And he had an aggressive pet Chihuahua. Which was lunging mightily at Harold.

"My turkeys," I said. Harold gallantly protected Lady Maude by puffing up to the size of a Rottweiler and standing in front of her.

Maybe Ben Franklin had been onto something when he proposed that the turkey should be the symbol of America instead of the eagle. These turkeys truly embodied the concept of American independence. They did their own thing and refused to sleep shut in the henhouse with the chick-ens. Instead, they perched on top of the chicken house, out in the cold. They could—and did—fly around the neighborhood. The third turkey, a Royal Palm like Maude, had winged off and was never seen again. I like to believe he ended up at the nature preserve at Lake Merritt, a few miles away, instead of as roadkill on the nearby freeway. There was an odd assortment at the sanctuary—a pelican with a goiter, a skinny chicken, and now, hopefully, a black-and-white-checked turkey strutting around, trying to mate with a duck.

Harold and Maude commonly took afternoon strolls down Martin Luther King Jr. Way. Though this is a regular thoroughfare for drug dealers, sex workers, and homeless men, the sight of two turkeys strutting down MLK nearly caused car accidents. The turkeys, on the other hand, didn't seem to mind the cars, the pigeons, or the sketchy pedestrians.

The turkeys were displaying a form of youthful behavior biologists call behavioral neoteny. Dogs, the animals that have been domesticated the longest by man, are considered neotenates: They don't have a species-specific sense of recognition, which means they will play with cats, goats, chickens, humans, or their own species. Dogs are also very curious, and they exhibit "juvenile care-soliciting behaviors" like begging for food. Other domesticated animals do the same thing.

It's thought that this is precisely how certain wild animals became domesticated. It wasn't human will, as many people believe, or that a baby animal of the domesticated species was found and thereafter raised among humans. According to Stephen Budiansky's argument in his influential book *The Covenant of the Wild: Why Animals Chose Domestication,* animals decided to be domesticated. Neotenates' behaviors, Budiansky argues, "would all have been powerful factors in inducing wolves, sheep, cattle, horses . . . to approach human encampments and to allow humans to approach them."

And turkeys probably did the same. Wild turkeys, native to the Americas, were most likely domesticated 2,500 years ago. Like most domesticated species, these birds chose to associate with humans—perhaps begging or following human encampments in South America. The ones who displayed the most curiosity, had the most open minds about different species, and could ask for help—like Harold and Maude—were the most successful. Eventually they were invited to live in human settlements. Their offspring were reared in captivity, fed and sheltered, ensuring an evolutionary future tied to man. It was a good bet.

As I explained my heritage-turkey pursuit the Hillbilly slowly nodded his head. When I started to babble on about how the turkeys were a product of thousands of years of domestication and how I was trying to reconnect to man's ancient contract with domesticated animals in order to rediscover my place in the natural world, he seemed to be looking at me in a different way. I realized the Hillbilly now had a name for me: the Hippie.

"Hey, you got a turkey over here," a man I didn't recognize yelled from the corner.

Harold and Maude had drifted down the 2-8.

"Can you . . . ?" I yelled. And he did. He herded them toward me by getting down into the age-old turkey-wrangling position: crouched low, arms open wide. I found myself making the exact same motions when I herded the turkeys, although no one had ever taught me this method. It's as if it was in our DNA, an embedded dance move.

The Hillbilly left, and I opened the gate and joined the passerby in convincing the turkeys to come back behind the fence. As we both moved slowly—arms open, hunched over—the man, who turned out to be from Tulsa, looked over at me and asked, "Where are we—Oakland or Oklahoma?" Our laughter persuaded the turkeys to retreat farther into the backyard. An ambulance then went by, and Harold gobbled out a warning.

Your chicken came into my house!" the Vietnamese guy across the street told me. He was very upset. "We had the door open, and it just came in!"

To me, this seemed like a minor nuisance. I mean, I once slept with a sick chicken. (Wrapped up in a towel. His name was Twitchy. He had a leg problem that never righted itself.)

When my neighbor saw I was nonplussed, he added, "It pooped in my house!"

Oh, dear.

"OK," I said, "I'll try to do something about that." But what? Give the chicken a talking to? Train it not to go outside our gates? In Seattle, our chickens roamed the streets with impunity. They had the run of our neighbors' backyards, and they sometimes walked down the sidewalk. But Seattle had more of a laid-back, suburban vibe. The houses weren't quite so close together, and the neighbors were less likely to be armed. None of our Seattle chickens had ever, as far as I knew, made it into someone's living space.

There was more going on than just impolite birds, though. Everyone was on edge that summer because of avian flu. It was killing people in Vietnam, where many of the people on our street had friends and family. Clucking my tongue, I went to the World Health Organization's Web site to get the skinny. "All evidence to date indicates that close contact with dead or sick birds is the principal source of human infection with the H5N1 virus," the WHO warned. "Especially risky behaviors identified include the slaughtering, defeathering, butchering and preparation for consumption of infected birds. In a few cases, exposure to chicken faeces when children played in an area frequented by free-ranging poultry is thought to have been the source of infection."

Dear lord, chicken poop could actually kill someone! It wasn't a good time to have twenty chickens in our backyard.

And yet I wasn't that worried. H5N1 hadn't reached American shores yet. My chickens couldn't pick up avian flu, and they couldn't give it to our neighbors, until the virus reached the United States. I promised our Vietnamese neighbor that I would get rid of all the poultry the moment H5N1 hit North America.

Every week, though, news sources threatened that avian flu was coming. Wild migrating birds, we were told, would bring the disease through Alaska and then to the mainland, where avian flu would kill countless birds and, eventually, maybe a human.

I bought some netting and stretched it over the chicken area so the birds couldn't get out. But within a few hours, they discovered an opening in my avian-flu shield and were back out in the lot, in the street, terrorizing the neighborhood.

A pandemic on the 2-8 seemed wildly fantastical, and yet I was starting to have my doubts. Especially after I read a *New York Times* article entitled "Avian Flu: The Uncertain Threat, Q and A: How Serious Is the Risk?"

Question three asked: If bird flu reached the United States, where would it appear? The answer: "Although health officials expect bird flu to reach the United States, it is impossible to predict where it may show up first, in part because there are several routes it could take. If it is carried by migrating birds, then it may appear first in Alaska or elsewhere along the West Coast."

I turned to question five, "How will I know if I have bird flu?" Symptoms, the article said, include flulike feelings: fever, headache, fatigue, aches and pains. But instead of getting better, the patient gets worse and ends up dying, in most cases from acute pneumonia.

Feeling a little congested, I sat at my table reading this news. I looked outside and saw the chickens marching around the lot. Chasing one another, pooping copiously. Suddenly they seemed sinister, out of control. Was a dozen eggs a day worth all this drama?

And so I became a pusher. A chicken pusher. Everyone in our neighborhood had a hustle, and this became mine. Chickens are, after all, the gateway urban-farm animal. I wanted others to join in the fun. "You'll get tons of eggs," I would whisper to my coworkers, "lots of fertilizer." No one in my neighborhood seemed interested, but Willow knew of some families. And then I posted an ad on Craigslist.

I had never seen such a parade of oddballs. A surfer guy who wanted to give his wife urban chickens as a tenth-anniversary gift. A chubby middle-schooler who translated for her Spanish-speaking dad. An eyeglasses-wearing gardening teacher who wanted some hens for her schoolyard. The teacher expressed interest in getting other animals, too, and so, like a dealer, I gave her my extra copy of *The Encyclopedia of Country Living*. I laughed when I imagined that soon the school would have bees and then a turkey, a few ducks.

I finally managed to whittle down the flock to a reasonable number, six, and the neighborhood gave a large collective sigh of relief.

CHAPTER SEVEN

❈

The automotive shop behind our backyard had a ten-foot-high fence, festooned with razor wire. Behind that fence lurked the shop's two large crime-stopping dogs. One day, Maude, being smaller than Harold and perhaps having an especially low species-specific sense of recognition, tried her luck clearing that fence in order to meet the dogs. Unfortunately, she succeeded.

I heard the barking, her shrieks, Harold's gobbles, and I came running. Feathers were literally flying. The cream-colored pit bull and the Rottweiler mix danced around Maude in the auto-shop yard. I scaled the fence.

Compared to the locals, I made a terrible fence climber. Every once in a while there is a car chase down MLK: squealing tires, police sirens, engines opening up. If the pursued car careens around our corner, it soon encounters a dead end: a schoolyard circled by a twenty-foot-tall fence. Not having options, the car thieves usually throw open the car door and sprint to the fence. I timed them once: five seconds to get to the top. The cops got out of their cars, lights flashing, and watched them climb away to freedom.

Now I got a startling reality check into what remarkable physical strength it took to scramble up a chain-link fence. The metal cut into my hands; my toes were jammed painfully into the small openings. Once near the top, I had to negotiate to an area without razor wire. My biceps quivered. I was a weak Spider-Woman. I yelled discouraging words to the dogs in my best stern voice: "No. Bad dogs! No."

Amazingly, instead of sinking their teeth into my ass, they backed away from Maude when I scaled down their side of the fence and landed in the auto-shop parking lot. On the asphalt, Maude lay torn and dead, her white and black feathers dotted with blood.

I stood in the middle of the parking lot and caught my breath. The Rottweiler looked up at me expectantly. If this had been Idaho, things would have been different. My parents' ranch dog, Zachary, wouldn't stop eating their chickens, and so one day, after another dead hen, my dad put a bullet in his head. That was country life for you.

I patted the Rottweiler's head without thinking. He didn't know what he had done. Maude's flight into their yard had to have been one of the highlights of their careers as junkyard dogs. There was only one way that scenario—turkey meets dogs—could have played out.

I was only a few feet from my backyard, but my house looked so different from here. Smaller. The gray paint was peeling; the stairs were scuffed and worn. I picked Maude up. Her eyes were closed; her throat was clotted with dark red blood. She was warm. Unlike Harold, she didn't have an impressive snood or wattles. From her head sprouted a few wispy black hairs. Though her body was small, she was dense and weighed more than a chicken. Her gray reptilian feet were scaled, but her toenails blushed with a bit of pink.

I tried to climb back with Maude in my hand. This was impossible, so I had to fling her limp body over the fence. On my trip back, the fence ripped my corduroy pants in the thigh and the crotch.

Harold discovered Maude before I reached our side of the fence. He was acting weird, puffing up and preening around her prone body. He was doing a mating dance, I think, as if he confused her current state with readiness to mate. Then he hit his head on the ground, making a strange thumping sound.

My eyes welled up at this curious spectacle. Harold mourned, and so did I. The injustice, the absurdity, of Maude's death upset me. But also, on a pragmatic note, as a burgeoning urban farmer, this was a serious setback. Maude was nearly full-grown, and she had been a lot of work, from teaching her to drink water, to cleaning up after her, to feeding her daily. Vast quantities of organic meat-bird feed and greens from the garden had disappeared down her now-ravaged gullet. I had practically risked my life to pick her weeds. And now she was dead.

Harold was making circles around Maude, some kind of turkey death dance that I had never heard about. It was amazing. But I felt awful. I was down to only one turkey.

Pushing Harold away after a few minutes of mourning, I scooped up Maude's body. Her tail feathers fell off and made a trail into the lot, where I carried her for her burial. I had hoped to celebrate Maude by serving her for Thanksgiving dinner. Instead, I dug her a grave under the apple tree.

As I laid Maude in the ground I recalled her generous spirit and remembered the time she pecked Harold's pendulous snood, mistaking it for a worm. How they'd slept together on the roof of the chicken house every night, cuddled like hobos under the pinkish glow of a streetlight.

I told Lana about Maude the next morning, and she cried. That it happened at the jaws of a dog—a fellow animal—didn't make it any better in her eyes. She just hated the fact that animals die.

I erased the 2 in "2 turkeys" on the chalkboard tally and made it a 1. I thought about my parents and their burned-down smokehouse. Bill shook his head. "What a waste," he said. He told me I should have killed her sooner. I could tell he was starting to seriously doubt my meat-bird plan.

Harold became a lonely turkey. His ancestors, brothers, and cousins lived in flocks. Though Harold had vaguely assimilated with the chickens, he always seemed to be looking for more solid company. So at night he scrambled onto our neighbors' roof to peer into their window and watch television with them.

As Thanksgiving neared he perched on our back porch to sleep next to the laundry line, emitting enormous turkey poops as he slumbered. In the morning, when I went out to feed the chickens, he greeted me like a lover, his tail up and feathers puffed. Two months to Thanksgiving and it was looking like Harold's end was going to be more of a mercy killing.

CHAPTER EIGHT

✦

I stood in a circle of light. I wore a hastily thrown-on pair of shorts and mismatched flip-flops. It was September and 3 a.m. I had a rusty shovel in my hand. Over and over, I chanted, "Don't." Clang. "Kill." Thud. "My." Wump. "Ducks." My chant was instructional but also moot, as the opossum I was beating had already killed a duck and a goose. Now he was going to die himself. He would take the lesson with him to hell.

I had awoken to the sound of the ducks quacking their heads off in the pen in the garden. I dashed downstairs and saw two fallen forms and a pair of reflection-lit green eyes inside the pen. Two ducks hunkered together near the pen entrance, quacking urgently, trying to get out. Their brother and the goose lay in the straw, not moving. Behind them, now trapped, their murderer, the opossum. Bill, who had followed me downstairs, wordlessly handed me a shovel.

Perhaps sensing a potentially painful situation, the opossum came at me and pushed his way out of the pen. I loathed how he moved—prehistoric, uncoordinated. His tail curled out behind him like a skeletal finger. I raised my shovel as he got closer and swung. With that one tap, he immediately fell into the grass. Lying there, he looked like a stuffed animal, or maybe a hairy taxidermied armadillo. One might have been tempted to think him deceased, but I knew the creature's patented skill: playing possum. If I stopped, he would eventually creep away, living another day to kill more of my farm animals. My weapon continued clanging against the marsupial's side.

Bill scooped up the surviving ducks and carried them upstairs. From the kitchen, he mounted a spotlight he used for fixing cars to help me to dispatch the murderer. The light also lit up my fallen animals—the bright white of the duck's feathers gleamed in the night. The goose slumped over in her cage, her neck broken.

My neighbor Neruda came outside and handed me her gun. I abandoned my shovel and tried to remember how to fire it. The gun was small, a purse gun, really, about the size of a butane torch that fancy people use to caramelize the top of crème brûlée. I had shot guns before, in gun-safety class at the middle school in the hick town where I grew up.

Neruda, in a fluffy pink terrycloth robe, shrugged and smiled in encouragement. Her daughter, Sophia, was asleep. Neruda must have needed the gun to feel safe in this neighborhood. That I could borrow a firearm like a cup of sugar sure felt neighborly. But in this case, it didn't seem right. With one eye on the opossum playing dead, I passed the purse gun back to her.

I picked up my weapon of choice again. If I were a movie gangster, I would've been the hit lady with a shovel in the back of my Cadillac. Channeling my rage, remembering the cuteness of my ducks, and the goose who would rest her head in my lap, I raised the shovel and came down on the opossum's neck. After a few thrusts—and, I admit it, grunts—head separated from body. I had my bloody revenge.

Somehow, this wasn't quite what I had imagined when I decided to expand my farm enterprise.

Only a few months ago, I had been signing for an air-hole-riddled box clutched by a mailman, anticipating liberation from the meat market. And now the mangled bodies of some members of the poultry package lay in a heap. How far I had fallen.

I had once wanted to be a naturalist, so I had read all the nature-loving books—*Pilgrim at Tinker Creek, Sand County Almanac,* even Barry Lopez's *Apologia,* in which he describes tenderly burying roadkill. Now there was a headless opossum in my garden, and I was seriously contemplating putting his head on a spike and posting it in the garden as a warning to all other predators.

Oakland's city code section 6.04.260 reads: "It shall be the duty of all persons having dead animals upon premises . . . to bury the same under at least four feet packed earth cover." If I didn't want to bury the opossum, I was supposed to call animal control and pay them to take the animal away for cremation. Otherwise, I would be guilty of an infraction. Forget the law. Even in my full-blown rage, I could see that the magnanimous thing to do would be to bury him.

Mr. Nguyen came outside smoking a cigarette and surveyed the damage. My yells had awoken him. Not that yells are uncommon in our neighborhood. There's one neighbor who shouts at her boyfriends and smashes dishes on the street late in the night. There are the shopping-cart crazies who smoke crack and then stand in the street yelling at their ghosts. But this was the first time in my three years in GhostTown that I had added my voice to the choir.

Mr. Nguyen clucked his tongue and shook his head when he saw the dead poultry. "Wow!" he said, gesturing at the opossum. I threw down my shovel. Mr. Nguyen picked up—no, cradled—the dead duck and gently set it out on the grass. He did the same with the goose. Then he disappeared back into his house with a wave.

This act of tenderness strangely inflamed my rage against the opossum.

Forget the spike. I would place the opossum in the middle of Martin Luther King Jr. Way, where he would be run over repeatedly. I shoveled him up.

I walked toward the main street, the opossum balanced on the end of my shovel. For a moment, I had the illogical fear that he would come back to life. But no, no, the head was definitely separated from the body.

Before I heaved the carcass into the street, I leaned against the bus stop to think. I felt jittery and wide awake. A few shadowy figures stood on the corner a few blocks away. What would they have thought had they looked my way: a perspiring white lady carrying a mangled corpse in a bloody shovel down MLK at three in the morning?

The abandoned building across the street loomed. Graffiti writers had been up there recently; an infamous tagger, Logo, had made his mark on the tallest part of the building. A lone car bobbed down the interstate.

The opossum must have lived there, in the slim greenbelt next to the highway. He eked out a living there on the margins, probably eating garbage and insects, nested up against the concrete. My ducks must have been a welcome snack for him, something fresh and delicious rather than the boring old garbage and grubs along the highway. Most likely this had been his first experience with a real farm animal. Like the junkyard dogs who killed Maude, this beast was just following instinct.

With the death of Maude, and now the duck and goose, I saw what a gamble it is to raise something that you care about. But the paradox was, I had planned to kill them myself, to eat. The dogs, the headless opossum—they were not the biggest killers. I was. Compared to what I had planned to do—roast the goose, confit the ducks, and truss the turkey—this opossum was a small-time player.

I tightened my grip on the shovel and looked down at the beast. His fur, I noticed, was a mixture of white and gray hairs. His paws were tiny and had sharp-looking claws. Small teeth peeked out from his mouth. His nose was pink, like that of a kitten.

Caught up in protecting my babies, I realized, I had become a savage. I was a little shocked to see the wildness in myself. That I could lose myself in human rage and commit this act of savage hate—I hadn't known I had it in me.

A few blocks away I could see a flickering-candle memorial. Churches in GhostTown had started a program called Stop the Violence. Bobby had even erected an instructional sign: STOP KILLIN' EACH OTHER. I suddenly felt very tired and sick of death.

My anger turned into exhaustion, I tossed the mangled opossum into the garbage can next to the bus stop. Take that, I thought, and went back into the garden, where I buried the duck and the goose under the apple tree next to Maude. I returned to bed just as the sun came up, a murderer.

A few weeks after the opossum incident, I went out to the garden to examine my watermelon. Yes, singular. The vines had unfurled throughout August and ran along the bed. Pale yellow flowers had come out. A watermelon must be visited by a pollinator eight times to ensure fertilization, so I had chastised my bees if I saw them working the cheap and easy fennel that chokes our parking strip. "Check out those melon flowers," I had urged them.

Deep in August, I had spied a swelling at the end of the vine. Just one. It had been a terribly cold Bay Area summer, and the rest of the plants were barren. Now the thing had ballooned, and its black stripes were beginning to show.

As I admired my sole, soon-to-be-harvested watermelon Jack Chan walked through the garden gate with a tall white man. The man wore a sunhat and had a bushy white beard. He carried a spray-paint can in one hand and some iron rods in the other. Somehow I had come to think of Jack Chan as my Emerson—a man not concerned with ownership, perhaps a transcendentalist who enjoyed communing with nature. Maybe he came to our garden and enjoyed its appearance, smiled beneficently at the towhee, a sweet but territorial sparrow that had come to live in the garden, along with the hummingbirds and fritillary butterflies.

Totally ignoring me, they began to mark property lines. I walked over to them.

They were standing smack-dab in the middle of one of my raised beds, whacking an iron post into the dirt.

"So, you're—" I started.

"Condos, right here. Three months," Chan, master of few words, said, turning to me.

"Oh," I said. My Emerson bubble popped. Chan was simply a real estate developer.

He smiled. "It looks great, thanks for cutting the weeds, but you will have to move it."

Then the white guy spray-painted the rod he had just placed, spritzing some of the leaves of my passion vine yellow.

I looked around the garden. What about my prized watermelon? According to a directory of heritage seeds called *The Seeds of Kokopelli,* the Saskatchewan promised to be "pale green with dark stripes. The seeds are black. The fruits are ovoid with very sweet-tasting cream colored flesh." Heirloom varietals often don't ship long distances well, the book explained, which makes them difficult to find in stores. In the case of the Cream of Saskatchewan, it has an explosive gene—if the fruit is knocked, it will split open. This seemed unbelievably sexy. What store could stock an exploding watermelon? It was now poised to be the last thing we harvested from the lot.

The empty duck pen lay near the blackberry bushes, where I had dismantled it after the opossum attack. Absurdly, I found myself relieved that the surviving ducks—they were living on the back stairs now—weren't here to see this. Suddenly, all the plants and trees I had regarded with delight seemed like a burden. I would have to dig them up? All that horse manure and dirt we had struggled to bring here, we'd have to get rid of it all?

I went upstairs, dread-filled, and watched Chan and his friend move around the lot casually, stepping in and out of beds, mindlessly crushing lettuces and herbs. They placed a total of four posts in the garden, sprayed them with yellow paint, and then left.

The bulldozers would arrive and level everything. They might even excavate the graves of my various dead animals. Inspired, perhaps they would name the condos Rotten Poultry Townhomes.

I heard Harold crying in the backyard. It's really a barking noise—three short yips. Harold had begun regularly flying to our neighbor's backyard, but he always had trouble getting back. After hours of dabbling, he would finally grow hungry and gobble and bark until I came to rescue him. This involved a ladder, a bucket, and furtive looks at my neighbor's back door.

I climbed up the ladder that I kept against the fence for just such instances. There he was, under the apple tree. Harold chirped and took a few steps in my direction. "Get your ass over here now," I ordered.

The woman of the house, a silent Vietnamese lady, came out at just this moment. She took in this ridiculous sight—me on the ladder, peeking over her fence and rebuking Harold the turkey—and flew into action.

Within five seconds, she had grabbed Harold (faster than I could ever catch him) and passed him over the fence to my waiting arms. Harold pretended to be a regal creature, used to being carried in the arms of a beautiful young woman. But once he returned to our side, it was back to the same old routine: chicken shit, circling flies, and loneliness.

I walked back to the lot. I saw that Chan and friend had posted NO TRES-PASSING signs on the gate to the garden. Did that mean everyone? Or were these signs a directive to me, their resident squatter?

I stood at the gate to the garden and peered in. The scarlet runner beans wound through the chain-link fence and were heavy with furry green beans. Huge squash rolled on vines. Malabar spinach, a heat-loving variety, twined up a trellis. Apples were ripening on the tree. Blood-red beet stems sprouted next to bushy basil plants. Eight varieties of tomatoes ripened in various beds. A stand of corn rustled in the corner. My presence, my influence, was evident all around me.

Thoreau, my fellow squat farmer, eventually ceded his bean field to the woodchucks. I would soon have to cede my garden to an urban farmer's most dreaded pest, the real estate developer.

From my window I would be able to watch the rewilding before the destruction. The tomatoes would turn red, burst open, ooze down their seeds in a slurry. The carrots would swell and split, send out a flower stalk, become fibrous. Armies of slugs and snails would slide across the wooden beds, tuck into the soil, and reproduce deliriously. The corn, neglected and unharvested, would crumple into the earth.

The Bermuda grass, my enemy, would creep over the whole lot in a ragged green mat. The oxalis would run rampant, and its flowers would light up the street with their lurid yellow. Eventually, fennel would sow itself in the raised beds. Then the boards would break apart. The propagation table would become covered with small sprouts, water glasses filling with rain. My garden would become feral, transforming back to what it had been three years earlier: a weed-choked, unloved, abandoned lot.

Imagining this place doomed, I wondered, Why hadn't I done more? Why hadn't I subsisted off this piece of verdant land? Why hadn't I sowed more, harvested more, given more to this piece of earth that I had grown to love?

Nature had been so good to me. The sun shone down. The rains came—and when they didn't, my socialist landlord paid the water bill. The worms and horses exuded nutrients. And the plants, which did all the work catching and using these gifts from nature, then produced a harvest. As a squat farmer, I had been a freeloader on many levels.

And yet, by doing this work, wasn't I simply repeating what humans have been doing for thousands of years? The seeds, these seeds that I had so carefully selected, were tangible proof of man's culture, of my culture, a continuation of a line. Even in this ghetto squat lot, I was cultivating human history. Watermelons from Africa. Squash from the Americas. Potatoes with a history in Peru. Radishes native to Asia but domesticated in Egypt. All now growing here in Oakland.

Standing near the fence, I realized that not only did I make the garden; it made me. I ate out of this place every day. I had become this garden—its air, water, soil. If I abandoned the lot, I would abandon myself. When Jack Chan told me no building—no permanent structures—only garden, did he realize that by building the soil, perhaps I was making something more permanent than he could have ever imagined?

I stared at the red letters: NO TRESPASSING. What does a sign in Ghost-Town mean anyway? Just as much as my signs urging people not to pick the garlic. In this forsaken place, NO TRESPASSING is merely a suggestion, a doomed hope. It might even be an invitation. I looked around for Chan and his sidekick. Then I pulled down the signs. I pretended that I was the wind and threw them into the street, and they became another piece of garbage blowing around the neighborhood.

CHAPTER NINE

It was time to rob the bees.

I walked out to the deck. The smell around the hives this time of year was both divine and fetid. Divine near the two top boxes, full of honey and pollen. The bees seal the cracks in the stacked boxes with a kind of yellow caulk called propolis—a sticky substance collected from tree sap and leaf buds—to keep out ants, drafts, and moisture. In the brood box, the deeper container through which the bees enter the hive after a day of foraging, the queen quietly lays all the eggs for the colony. In the fall, her production slows. The colony spends cold nights huddled in a ball in order to keep one another, but mostly the queen, warm. The smell from these stacked boxes is ambrosial—earthy pine, beeswax, and sweetness.

Not so sweet smelling is the quagmire of dead bees piled up outside the hive at the end of a season. During prime nectar-gathering time, up to one hundred bees a day die inside a nest, *The ABC and XYZ of Bee Culture* told me. The corpses are "carried away from the colony in the mandibles" of a caste of bees known as the undertakers, which recognize the dead by a chemical odor. It looked as if my undertaker bees just tossed the dead over the edge of the hive. Since it was on a deck in the middle of a city, the corpses didn't gently rot into the soil or get blown away by the wind. They simply rotted on the hot roof—and the resultant reek was piercing.

On a sunny October day, Bill and I stood on the deck, taking in these odors. After cracking the propolis caulk around the bee boxes, he hefted the uppermost box, or super, and I slipped the bee escape—a beekeeping tool that consists of a wooden box with a pattern of openings—underneath it.

The bees in the now-sequestered honey super could leave through the bee escape's little tunnels, but they couldn't get back up. It would take about twenty-four hours for all the bees to empty out of the hive's honey store-house. Most commercial operations use blowers or noxious fumes to drive the bees out. The bee escape seemed less offensive, and sort of fun, like a practical joke played on the bees.

The next day, my friend Joel, an Oakland public school teacher and former Deadhead, came over with his children, Jackson, ten, and Margaret, eight. They spilled out of the car, and Joel bounded up the stairs of our apartment. Joel and kids were interested in keeping bees and wanted to get the full experience before they got their own hives.

Margaret couldn't believe how cramped our apartment was.

"There's something everywhere!" she said.

It was true that, over the year, things had gotten pretty messy. We had bee boxes stacked up against one wall. The extractor had arrived in the mail a few days earlier and dominated one corner. I had acquired various farm implements—shovels, rakes, pruners—which hung in our laundry room. The pantry was overflowing with canned tomatoes and pickles. The kitchen table was so stacked with books about building cob ovens and keeping goats that there wasn't much room for anything else. The fall crop of lettuces was germinating on the windowsills. I had acquired more fly strips to battle the increasing numbers. The two survivor ducks were living on the back stairs.

The house where Margaret lived was very neat and orderly, and there were rules about what belonged inside and what belonged outside. I gave that up when I started urban farming. There's a principle in intensive urban farming called stacking functions. I told Margaret about the concept.

"See, these extra bee supers are also our coffee table," I said as I showed her around. "And the deck is a garden and a bee yard." I felt like Pippi Long-stocking giving a tour of Villa Villekulla.

Joel and Bill lifted the honey super off the bee escape and ran into the house before the bees could figure out what had happened. The super was heavy with honey. Inside the fragrant, beeless box were ten frames, crammed in like library books. When we pulled one up, it was revealed to be a big chunk of sealed honeycomb, like a slab of gold within the frame.

We lifted out each frame, scraped off the wax-capped honey with a serrated knife, and spun it in our shiny new stainless-steel honey extractor. Before the invention of the extractor, in 1865 by a beekeeper named Major Hruschka, people mashed the entire honeycomb and strained the honey from the wax, which took many days and attracted pests. Hruschka realized that it would be easy to remove the honey from the sturdy frames by using centrifugal force. He concocted a spinning device in which the frames could be mounted and the honey would fly out.

We made the kids give the hand-cranked extractor a spin. Centrifugal force launched the honey out of the comb and onto the stainless-steel wall of the extractor. Then the honey dripped down and collected at the bottom, where a spigot opened to let it pour out. Commercial beekeepers use plug-in knives, automated de-cappers, and motorized extractors, and they heat and filter their honey. Instead, Joel and family steadied our extractor, which had a tendency to keel over, cranked it as hard as they could, then let the honey drizzle out into a few quart-size mason jars.

We were all sticky with honey and buzzed from licking our fingers and chewing on the leftover wax, which reminded Margaret of chewing gum. We extracted eight quarts of honey in less than an hour. When we lived in Seattle, it took days to get the honey out. Our new machine was impressive indeed.

"If you get bees, you're welcome to borrow this," I told Joel.

I could see it in his eyes. Hear it in the giddy laughter of his children. Smell it in the heady liquor splattering on the sides of the steel tank as we spun, the scent wafting up into the nostrils of the person cranking the handle. Soon, I predicted, Joel's house would be as big a disaster as mine.

He and the kids took home jars of honey. I brought one to the chubby elder monk who lived across the street in the monastery. When I handed it to him and pointed at the beehive on our deck by way of explanation, he gazed up at the hive, eyebrows furrowed. After a moment, he seemed to understand. He smiled and snuck the jar deep into his yellow robes. Without a sound, he turned back into the temple and disappeared.

After cleaning up, we returned the now-empty honey super to the hive on the deck. The bees simply went to work salvaging any remaining honey and wax and cleaned up the mostly empty frames. I felt a little niggle of guilt. Bees have never been truly domesticated—they are not tame and didn't really depend on me to live, as Harold and the chickens did. My bees, in fact, were of the same genetic stock as wild bees, bees who make their homes in trees. All we could do was offer them a home and hope it was enough to convince them to stay.

I couldn't help but relate to the bees. Their box, like our lot, was a temporary thing—a home for the moment. At any time, the bees could decide to leave, and wherever they relocated, they would make wax and store honey. I realized that that would be our fate, too: wherever we ended up going after the lot was bulldozed, we would build a garden, keep chickens, set up beehives. It's just what we do.

As for raising meat birds, Bill was still unconvinced. I was too attached to the survivor ducks to ever kill and eat them. Harold was my last chance to prove my box of poultry hadn't been a dismal failure.

Bobby started going barefoot most of the time, which is dangerous in our neighborhood, where the streets and sidewalks almost uniformly glitter with broken glass. When I told him about Jack Chan's plans, he said, "Condos?" and looked dazed. "Here?" He looked at the abandoned brick building across the street. Then he shuffled off to sweep the street, which he did nightly.

I picked my third—and presumably last—crop of tomatoes. The lettuce started to taste bitter. I expected to see bulldozers any day now. My watermelon, coveted thing, was growing despite the setback. It hadn't heard the news of the impending destruction. Worried that I might harvest it too late or, worse, too soon, I asked around for advice on how to tell if a watermelon is ripe. No one in the city could help us.

Bill and I went to see some farmer friends who live in Mendocino County, about three hours north of Oakland. They had draft horses and forty gorgeous acres of row crops, and I felt a little self-conscious when I told them

I was an urban farmer. I was micro-scale compared to what they were doing. I knew that they resented having to drive into the city to sell their vegetables, but they had to go where the population center was. Growing food in the city cut out that step.

Over a dinner of their farm's roast beef and potatoes, the youngest son of our farmer friends—a strapping boy of seventeen who already knew how to build a barn and castrate a pig—told us how to tell if a watermelon is ripe. You look at the spot where the melon has been lying on the ground, he told me. If it's a pale, pale yellow, it's ready.

After a restful day, we drove back to Oakland, the watermelon on my mind. Coming back to the city from the country takes a few hours of adjustment for me. I'll wander off and not lock the car door, for instance. A farmer once said that the only time you have to lock your car in the country is during zucchini season. If you don't, you'll wind up with a passenger seat full of oversized vegetables. But if you value your car battery in the ghetto, you better lock your door.

Feeling relaxed and open, ready to apply the farmer-approved method on my melon, I walked toward the scrambling watermelon vine. It was Indian summer, still warm during the day, but the nights had become chilly, and it was getting dark sooner. I wondered why construction on the condos in the lot hadn't started. It would be rainy season soon.

I had to look under a few leaves before my brain could accept the truth: The watermelon had vanished. Wrong bed? No, there's the vine. I gingerly traced the splay of the plant, from the mound where it had first emerged to the tip where the fruit had been. Had been. Where the watermelon had been lying, its weight (five pounds?) had pressed down into the soil and made a depression. The divot was the only thing left. I crept upstairs and cried.

Motherfuckers.

There were suspects, of course. Lou, the collards harvester, being number 1. But there were so many. Some of the corn had gone missing recently, too. I wondered if people in the neighborhood could sense that the garden was doomed. That the yellow spray-painted property lines had marked the garden as weak or crippled. As when wolves take down a weak member of a herd, the garden became a target. But the callousness, to take my one and only watermelon, even if it was doomed, boggled my mind.

And I wailed when I remembered that I wouldn't be able to save the seed, pass it on to someone else's garden once mine was bulldozed. Worst of all, I would never know how the Cream of Saskatchewan tasted. I couldn't just go buy one.

I wanted to find the culprit, strap him in a chair, and ask him a few questions. Like: Did it shatter when you cut into it? Was it creamy on the inside? Sweet tasting? The best goddamn watermelon in the world? This fucked-up neighborhood. Bobby? Did he do it? That son of a bitch. I looked out my window. Bobby was sitting barefoot in his lawn chair in the middle of the street.

I burst out laughing. Let them build condos here. I dare them. Lunatic Landing. Crackhead Townhouses. Broken Window Live/Work Lofts.

CHAPTER TEN

✳

Mindful of Thanksgiving, Lana invited Harold to live out his days in her warehouse.

I demurred. I wasn't a hobbyist with a pet. I was a farmer with a turkey on my hands whose feed-to-weight ratio had reached a plateau. I had fed, groomed, defended, loved, spoiled, *named* this animal. And now it was time to harvest him.

On the big day, I woke up late and filled with dread. I puttered around the kitchen. I boiled several pots of water and poured them into a large metal vat outside. Read the paper. Found the ax. Tied up my hair in braids. Finished a novel I had meant to return to the library. Sharpened the ax. Put on my dirtiest clothes. Cleaned out the fridge. Set up a chopping block. Boiled more water. Then, as the afternoon sun streaked the November sky orange, I realized that I hadn't seen Harold around.

Not in the backyard. Not on his usual perch on the stairs. Not in the neighbor's backyard. Harold was supposed to die at my hands but was nowhere to be found. I'll admit it: I was relieved.

The previous night, I had been a wreck. The facts were before me. This Thanksgiving, I was not going to be a consumer buying a free-range turkey. This year I was a producer. Against all odds, one of my turkeys had made it to harvest weight, and I better kill him now before someone else did. But first I had to figure out how to do it.

Seated at my kitchen table after dinner that night, I had flipped through *The Encyclopedia of Country Living*'s newsprint pages until I came to the poultry chapter. Carla Emery writes, "I don't think much of people who say they like to eat meat but go 'ick' at the sight of a bleeding animal. Doing our own killing, cleanly and humanely, teaches us humility and reminds us of our interdependence with other species." I had nodded my head and quickly turned to the section titled "Killing a Turkey."

Emery's words of wisdom:

"The butchering process with a turkey is basically the same as that with a chicken except that your bird is approximately 5 times bigger."

"First, catch the bird and tie its legs."

"The turkey may then be beheaded with an ax (a 2-person job, one to hold the turkey and one to chop)."

Lying in bed, I had worried that I would botch the execution, that Harold would feel pain, that his feathers wouldn't come off, that I wouldn't be able to clean the meat properly. So I visualized, I rehearsed. First, ax to neck, then bleeding, then defeathering, then cleaning. I mumbled a macabre lullaby before falling asleep.

But now I wouldn't get the chance to practice what I had memorized. Harold was smarter than I'd given him credit for, I thought. He knew what was afoot and simply flew away.

Bill had to work, so Joel had come over to help. Eager to teach his children about where food came from, he brought his ten-year-old son, Jackson. The three of us stood in the garden, the sound of the traffic from 980 roaring by, a vat of steaming water and a sharpened ax nearby. We wondered what to do.

Then I spotted Harold. He was perched on a low fence in the garden, a place he had never ventured before, watching us. "There he is!" I yelled. It was weird, as if he had known. Harold stood and adjusted his perch. I picked him up; at twenty-eight pounds he was quite an armful, but he liked being held and didn't struggle.

Jackson smoothed Harold's iridescent feathers and looked with wonder at the giant snood that dangled over his beak. The turkey's voluminous wattles billowed below his neck like an old man's jowls. I told Jackson about Harold's life over the past six months, his adventures, his grief over Maude, and his future: on our Thanksgiving table. Jackson's blue eyes, hidden behind a giant pair of glasses, darted around. Then he blurted out, "I don't want to see *it*."

So I tucked Harold under my arm, took Jackson over to a weedy patch where the watermelon had once grown, and showed him how to pull weeds. Then I faced a task that, hell, even I didn't want to do.

Harold and I, there in that squat lot, embodied the latest endpoint of centuries of mutual dependence. The only reason Harold existed at all was that he and his ancestors had made a Faustian bargain with humans: guaranteed food, shelter, and the opportunity to pass on their genes in exchange, eventually, for their life. Paradoxically, to be killed was a way of life for Harold.

Though Harold had become a postmodern turkey—named, citified— his death, when viewed in the context of his species, was part of being a turkey. As the naturalist Stephen Budiansky points out in *The Covenant of the Wild,* "All of nature's strategies for the survival of a species, strategies which include domestication, include suffering and death of individual members of that species." Harold's mother, some breeding hen at Murray McMurray, lives on. In an indirect way, Harold's mother and I were cooperating so that we could both survive.

Of course, we meat-eating city dwellers don't have to kill something to survive. We merely go to the store with some cash in hand. How many people would eat meat if they had to kill it themselves? This was the question I had pondered for six months as I watched Harold grow from a puffy chick into a full-grown turkey. I eat meat, I like eating meat, it is part of my culture and, some might argue, my heritage as a human being. While Harold had to die, I had to kill.

At the grocery counter or farmer's market stall, the cost of the meat I bought factored in the cost of the bird's life—feed, housing, transportation to market. A small portion of that cost included a kill fee. I had been comfortable allowing someone else to be my executioner. And suddenly, all the meat I bought, even though I had considered it expensive at the time, seemed underpriced.

We burned a little tobacco in an oyster shell, as a new-agey friend had recommended. She said it was a Native American tradition that showed the animal's spirit which way was up. The !Kung people, a hunter-gatherer tribe in Africa, ask forgiveness of an animal's spirit. Budiansky tells us, "They don't pretend there is no ethical cost, or guilt even, inherent in the act of killing the animal." With this in mind, I whispered into Harold's ear my thanks and asked for forgiveness.

Although I usually call myself an atheist, a lonely universe offers little comfort to a person confronting death. I thought of my father, a voracious hunter and fisherman who, even after my mom left him, never came back from the land. He had tried, in his subtle way, from a great distance, to instill in me some of his beliefs about nature as a god of sorts. In my life as a city dweller, though, pantheism had mostly eluded me. But to hold Harold, this amazing living creature, and to know that his life force would be transferred to me in the form of food, felt sacred.

I stroked his warm, warty head. The folds of skin were soft and pliable, punctuated with small wayward hairs. I could feel his heart beating, slowly.

It's true that eating meat springs from a violent act, that in that way meat is like murder. But it is not an act filled with hate. I had murdered the opossum. But Harold I had raised in a loving, compassionate way, with good food, sunshine, and plenty of exercise. Harold had had a good life, and now he would have a good death—quick and painless—at the hands of someone he knew, in a familiar place.

"OK, Joel. Ready?" I asked.

Mrs. Nguyen, who is the only practicing Buddhist in the house, perhaps sensing our murderous plan, dropped her blinds loudly.

Joel looked nervous but steadfast; I could depend on him. After all my stalling, it was almost dark by the time we finally laid Harold's neck across the chopping block. For his part, Harold seemed resigned, bored, as if this scene had played itself out a thousand times before. I swung the ax. I swung again. Harold had a really big neck.

Muslim tradition says one must look an animal in the eye until its soul departs. I was satisfied that Harold and I had had a sufficient dialogue. He gobbled once, a warning sound that he regularly made. It made me a little sad to think that in the moment of his death he might have been scared.

It was a solemn moment. I hefted what was once Harold to a bucket in order to bleed him out. Though headless, he thrashed mightily, and the bloody neck stump pointed accusatorily toward me. I felt relieved, a little ashamed but giddy.

After his body stopped thrashing, I lowered it into a large pail of steaming hot water. His tremendous girth displaced some of the water, which flowed over the edge and into the garden. After fully submerging the body for a few seconds, I pulled it back out into the air. Joel and I sat down and plucked the feathers like a couple of old farmhands. They slipped off in clumps.

I could finally take a deep breath. I looked around the garden—only a few fall crops had been put in. Bill and I were expecting the bulldozers any day. We had become glum and unmotivated gardeners. The billboard that overlooked the lot had a public service ad warning against sexual predators. BART clattered by. A few loud teenagers shuffled down the sidewalk but didn't look in. Killing Harold, we thought, was one of the last things we would do in this garden.

Joel and I made small talk, discussed how the death had gone, how beautiful Harold's skin looked beneath the feathers, how tasty this homegrown turkey would be. As the feathers flew and more and more skin was revealed Harold started to resemble a turkey you buy at a store, wrapped in plastic and defrosted in the sink.

Looks were deceiving, however, as I still had to clean out the guts. Joel and Jackson did not want to stay for that part. I waved goodbye to them and took Harold upstairs. Alone in the kitchen, a place where I had cooked innumerable meaty meals, I got out the *Encyclopedia* again and turned to the "Cleaning the Bird" section.

Harold's body was still warm when I laid the carcass across a paper-lined table and made the first incision.

First I removed his crop, the baggy sac near his neck, filled with grain and greens he had eaten that morning that hadn't made it to the gizzard for digestion. I identified the trachea. Then, after a few precise cuts near the pooper, coached by Carla Emery, I eased out most of Harold's viscera with one steady pull. The curlicuing small intestine, the healthy dark liver, the pert heart and foamy lungs. The gizzard was round and covered in silvery skin. Curious, I made a slit down its tough side and examined the contents. Along the muscle walls was a green and yellow paste. It looked like wasabi but smelled like swamp. Within the goo were a few pebbles and a lot of smooth pieces of glass. Harold had been an urban turkey through and through.

I buried the inedibles in the compost. Everything else I took from Harold I used. I chopped off the feet—I knew a punk who wanted those, even though she said she wouldn't eat a bite of him. Her girlfriend wanted his wing tips, the dark-feathered ends of his three-foot-long wings. She said she'd use them for a costume. Their dog got his swampy gizzard. The enormous turkey neck was ringed with yellow fat, which boiled up into a rich gravy.

By the end of the process, *The Encyclopedia of Country Living*'s pages were marked with blood. And brining in the fridge was a heritage-breed turkey that I had raised from a day-old chick. The poultry package—bought with a credit card and priority-overnighted—had turned me into a farmer.

Now that my work was done, I had to trade in my straw farm hat for the paper one of a chef. God sends meat, and the devil sends cooks, as the proverb goes. As I morphed from farmer to gourmet—the fussier, snobbier element of food production—I worried about how my turkey would taste. On the advice of Carla Emery, I let him "rest" in the fridge for a couple of days so the meat would be more tender. If his flavor was off, his entire life would have been wasted. The burden felt heavy, and as Thanksgiving approached, I fretted more than usual. While this was a heavy load to carry, it was exactly what I had hoped for: meat had become a sacrifice—precious, not a casual dalliance.

On the big day, I put the turkey in the sink and trained a light on the body. I picked away all the little feathers and tweezed the wayward hairs. I made a few strategic cuts in his fatty skin and slipped in garlic cloves, herbs, and butter. Then I anointed his whole body with olive oil and salt. Once I placed the turkey in the oven, a wake became a dinner party.

I invited Mr. Nguyen over for dinner, but he shyly demurred. Willow arrived with a pot of stewed greens. Bill anxiously awaited the results of the meat experiment. Joel had a family engagement and couldn't make it. By dinnertime, ten guests had arrived. We toasted Harold before eating. I had snuck a sample before serving, so I already knew how good the turkey tasted. His thigh and leg meat were the color of milk chocolate. His flesh was perfectly moist, buttery and savory. His skin crackled. Everyone agreed—each bite was special. Bill gnawed on a drumstick and closed his eyes with pleasure. It was the best turkey he had ever had.

By meal's end I was uncommonly satisfied and full. I looked at my dear friends seated around the table and felt humble and grateful to have nourished them. Then I piled a plate with dark and white meat and went downstairs. I paused at Mr. Nguyen's door, then knocked. I wanted to bring him food and proudly say, This is how it used to be done in America. The plate of turkey was a tasty piece of evidence of an earlier, very different time. I thought of my parents while I stood there, about how they had once salvaged their turkey out of the charred remains of a smokehouse. I had essentially done the same thing.

Mr. Nguyen giggled when he saw the plate in my hands, just like when we had peered into the peeping box of baby birds. "Yours?" he asked as he took it with a grin. I nodded proudly. Behind him, in his living room, I saw his family's altar. It glowed with red lights and incense.

Back upstairs, the ten guests and I polished off the rest of Harold. In the end, I was left with a carcass.

✵

CHAPTER ELEVEN

Winter arrived. The hens got wet. The crops grew slowly. Bobby put a tarp over the roof of his car.

Then, in deep winter, something happened to me that made living in GhostTown suddenly seem a lot less fun. January in Oakland meant cold nights and lots of rain, so at 5:30 p.m., as I rode my bike home from work at the plant nursery, it was already dark. A light rain misted my glasses.

I saw a pack of twenty or so kids hanging out on the corner a few blocks from the park where I picked weeds for the chickens. Instead of avoiding the kids, I rode right by them. Some of them were on bikes, too. I hadn't thought much about it.

One time on this same street I had encountered a bunch of teenagers playing football. Since it had been night, I couldn't see the ball, just twelve six-foot-tall teenagers running toward me. I had nearly pissed my pants with fear. Then the football had bounced at my feet, and I had laughed at my wildly beating heart.

This pack, unfortunately, wasn't playing a game. One of them kicked my tire. Then a circle of kids ranging in age from twelve to sixteen formed around me. The kid who had kicked my tire, a thirteen-year-old in a puffy parka, suddenly had a gun in his hand. I couldn't tell if it was real or not.

I dismounted from my bike. Lana's word, "babies," flashed through my mind. This was the exact thing I had always feared would happen. It was why I had at first been reluctant to walk around our neighborhood gathering weeds for the chickens. By now my fear had been erased, though, and I felt like I was part of this place. I liked to think I understood our 'hood's dynamic and my place in that dynamic as the resident, slightly batty, urban farmer.

"What do you think the police will do to you if they see you with a gun?"
I asked the kid. I had been here long enough to know that I wasn't prey and
that this kid was not a predator.

He didn't say anything, just fingered the metal thing and scowled. I put
down my kickstand and began a strange oration, which at its heart was moth-
erly. I had killed an opossum with a shovel and axed a turkey with my bare
hands—did he understand what kind of crazy bitch he was dealing with?

"I'll tell you. The cops will kill you." A dark car eased by a few blocks
away. "And if they don't, someone else will. Some other gang member. You
have given them a great excuse to kill you. To just . . . execute you."

I swallowed. I was in tears, but my heart was swelling with love for this
kid. "You have to be careful," I told him, thinking about the R.I.P. candle
altars. "I care about you. Please be careful or you'll end up dead."

The kid made a face and walked away from me. I climbed back on my
bike. The circle broke up, and the kids let me go by. A few of the group followed me and told me their friend was crazy. I pedaled slowly with them.
"You have to help your friend," I said, "or he'll be dead soon."

When I got home, Bill was in the tub reading a book about Bob Dylan.
The bathroom was steamed up and foggy. From the wrinkles on his wide
feet, it looked as if he had been soaking for several hours.

"I just almost got mugged," I said, only realizing it then. I sat down on
the toilet lid and started to shake.

Bill put his book down. "What happened?"

I told him about the kids, the gun, my strange lecture. I cried in the
steamy bathroom about the stupidity and injustice in this world, the cycles
of violence that seem like they will never end, and my inability to change
anything.

"Do you want to move?" Bill asked, looking concerned.

I looked at my hands. My nails were ragged and my fingers were lined
with dirt from working at the nursery. What if that gun had been real? I
allowed myself to think. What if I had been shot for my bicycle? I knew bad
shit happened in lots of places, but I had a choice: I didn't have to live in this
place with so many problems.

But I couldn't imagine any other neighborhood where I could have turkeys and chickens, bees and the squat garden. That lot, that verdant place,
destined to become condos. If I lost the lot, if the bulldozers came, that
would force us to move on. Maybe Bill and I would relocate to North Oakland, which had less random gunfire and fewer muggings.

If my fate depended on the lot, I had to find out the truth. The next day
I called the City of Oakland building-permit office to find out about the
soon-to-be condos. In five minutes, I knew everything about the lot. Chan
had bought it for $40,000 just before we moved in. His building permit had
been rejected.

"Hey, the rainy season is here," the deep-voiced permit officer told me.
"There's no way he's going to build this year."

I mentioned that our neighborhood was a bit rough. How could they sell
condos here? I asked him, just talking out loud.

"Yeah, no one's going to finance that," he said.

And I hung up the phone, knowing that my garden had a stay of execution.

I also recognized a paradox: As long as our neighborhood stayed messed
up, I could have my squat garden and my menagerie. Bobby could keep
his improvised home on the 2-8. I might still lose some of my produce, like
the coveted watermelon, but maybe that had been an offering in return for
all our strange blessings. I made a pledge to grow the garden even bigger,
to raise more animals, to keep going as long as we could on this chunk of
squatted land.

Even while I plotted for the next spring, I felt a little sorry for Jack Chan.
As a real estate developer, he'd had his plan squelched. Though his loss was
my gain, I could empathize. A bold plan—in my case, a tenuous garden
and a box of live poultry; in his, a lot in the ghetto transformed into luxury
condos—tends to bring with it a great deal of heartache.

PART II

RABBIT

CHAPTER TWELVE

❋

I'll admit it: I missed Harold the turkey. At least, the idea of Harold. It had been a delight to have his unapologetic turkeyness strutting all over the neighborhood. I did not, however, miss his incessant gobbling, having to clean up his enormous turkey poops, or the constant uneasy feeling that he might die at the jaws of a junkyard dog or take off for Lake Merritt before I could enjoy him for Thanksgiving dinner.

Having slain that last fear (literally) but still a little exhausted from my meaty experiment, I felt a touch of worry as I climbed into my station wagon and steered it north into the Berkeley hills. It was early spring, and the air, even in the ghetto, was fresh with misty rain and possibility. My wipers didn't work, so I had to rub a rag across the windshield at red lights.

A woman I knew vaguely, Nico, had recently acquired some rabbits. I had met her while working at an alternative-fueling station in (of course) Berkeley. After paying for her vegetable-oil-based fuel, she had handed me a business card, which was a seed packet of sweet peas with her name and number written across it. We had hung out a few times, mostly at her boyfriend's farm in Pescadero.

I turned east on Dwight and began to wind my way up a forested hillside near the UC Berkeley campus. The sharp mentholated cat-pee odor of a eucalyptus forest wafted into my car and combined with its newly acquired barnyardy reek. Last week, delirious with our annual spring gardening fever, Bill and I hadn't been able to borrow a truck to get our usual loads of manure. So we had improvised: We lined the back of the station wagon with a tarp, drove to the stables, and loaded up. Of course, the tarp could only do so much. Many a clod of horse manure had broken free and now lurked in the car's surprising number of crevices.

Following Nico's directions, I crested a steep hill and pulled into her driveway. She lived in a rustic cabin, a wooden cottage nestled deep in the forest. At her door, I noticed that chaos was clearly unfolding. Stacks of boxes cluttered the stairs. Bottles and jars of various homemade ketchups and jams stood in front of the garage. One cracked jar gently wept out its sticky contents.

"Hello?" I called up through the cabin's open door.

Nico clambered into view. "Come in, come in." Her frazzled blond hair looked more disheveled than ever. A few years ago, Nico had relocated from Boston to finally finish college at UC Berkeley. She hadn't been successful so far because she was always getting distracted with some project or other.

Something skittered across her kitchen floor.

"Bunnies!" Nico shouted when she saw my eyes following the shadow.

There were four of them—two with white and brown spots, one pure white, and one solid brown—milling around the couch.

"The woman I bought them from," Nico said, offering me a home-made pickle from a murky mason jar, "lived entirely off a quarter of an acre of land."

"Really?" I said. "Eating rabbits?" I didn't know much about rabbit tending, except that my back-to-the-land parents had once raised them for meat.

Nico's plan was grander than mere survival; she had high-end dining in her sights. Rabbit had recently been showing up on the menus of fancy restaurants, and Nico, always a dabbler looking for a new project, bought three young females and a solid brown buck named Simon with the idea that she would sell their offspring to these restaurants. She couldn't have been doing it for the money. Though her hair was matted and her clothes tattered, she was a trustafarian, a common species of Berkeley resident. Her dad was worth millions. I think she just wanted to be a farmer, even a small-scale one.

Once the rabbits were of breeding age—around eight months old—Nico planned to mate them, and after two months she'd harvest the offspring. It didn't sound like fun to me, but Nico was always trying something different. In the short time I had known her, she had started a landscaping business, then gotten distracted by starting a catering company, and now was getting distracted by the rabbit business. The constant dabbling was a pattern, I had noticed, shared by many a trustafarian.

I reached over to pet a particularly soft-looking female. She made a screeching grunt and hid under the couch.

"These bunnies," Nico said, "are like mini grass-fed cows!" Grass-fed beef was all the rage in the Bay Area at the time. Guilty carnivores had realized that the soy and corn fed to beef cattle could feed starving people instead. But if the cows ate pasture grass, inedible to us and starving people in Africa, then we could have our steak *and* the moral high ground.

"Sounds good, Nic," I said. Yes, her idea was solid. But why had she invited me over? And what was up with the boxes?

It turned out there was one holdup in Nico's rabbit venture.

"I'll be gone for three months at Ballymaloe," she said, slapping a baseball hat on top of her hair. Ballymaloe, she explained, was a cooking school on an organic farm in Ireland. Finishing college, it seemed, could wait another semester. In Ireland Nico hoped to learn the fine arts of preserving food, making cheeses, and, no doubt, cooking a rabbit. She took a bite of pickle. "Six at the max—there's a Swiss goat farm that I might go to. . . ." She got a dreamy expression on her face. Nico loved, almost worshipped, farmers.

Then she informed me that her bunny-sitter had flaked out. So would I . . . ?

I hesitated. It was as if I had found myself at a party and been offered a new drug. It was one that I had never intended to try, but now, with the offer on the table, I started to remember all the good things I had heard about it.

Namely, that rabbit manure is manna for the garden. Their turds are like those of Lana's guinea pig—round, vegetarian, and nutrient rich. Chicken manure has too much nitrogen to apply directly to a garden; it first has to be composted, or it will burn crops with its scorching nitrogen load. Rabbit poo, on the other hand, is almost like compost from the moment it passes by the bunny's furry tail. Its nutrient levels are the perfect harmony of nitrogen, potassium, and phosphate.

I chewed on my pickle and looked out Nico's window. She, sensing that I needed time to process, fiddled with her belongings. I could see the San Francisco Bay from her window. Nico had once come to my house and refused to get out of her car, she was so scared of my neighborhood. It's funny, most trust-fund kids would be blowing their wad on fast cars and cocaine, and here was Nico, geeking out on preserving food and raising rabbits.

"You know that whatever you breed, you can eat, too, right?" Nico said, folding a sweater, sweetening the deal.

I nodded my head. She had heard about the turkey and knew about my tendencies. I had actually been researching a miniature breed of beef cattle when she had called. My plan was to put the mini steer out on the abandoned-schoolyard field. Alas, a mini cow was not in my price range. The grass-fed rabbit concept was definitely more feasible. Besides, the peer pressure was excruciating. I sighed and finally said yes.

Nico let out a whoop and gave me a hug. We packed the bunnies into a cardboard box and loaded them into the horse-manure-scented station wagon.

"This is going to be great," Nico said as she threw a bag of rabbit food into the backseat. Then she dashed back inside to finish packing.

As I drove through Berkeley and into Oakland, I looked down to see that the brown bunny, Simon, had squirmed free of the box and was now perched in the passenger seat. A cold March wind blew into the car as I dangled my arm outside to wipe off the windshield. I sighed. This—a bunny riding shotgun—had not been part of my urban-farming plan. There were many things that I would have to learn about rabbits. And there was no doubt that they, like Harold, would present their own unique challenges.

The most obvious problem was that in order to get that grass-fed meat, I was going to have to kill one of Simon's babies. Probably an adorable fluffy baby. Simon found a rip in the car seat and started to paw at the frayed fabric, as if he wanted to burrow into it.

I wasn't sure how I felt about killing a furry mammal. Normally I would not have gone out of my way to start raising rabbits. Yet it wasn't entirely out of my range of experiences. I had eaten a lot of rabbit while growing up on my parents' ranch. Tastes, as they say—and I recall—like chicken.

At home, I put Simon back in the box and carried the rabbits into the backyard.

Bill was tinkering with an engine in the shed.

"What is that?" he asked.

"Nico's rabbits," I said, trying to sound cheerful. I put the box in what had been the chicken run, a wire-enclosed pen connected to the chicken house. The chickens were going to have to learn how to share. The rabbits peered over the cardboard flaps of the box, then hopped out.

"So what are you going to do with them?" Bill asked, warily watching the bunnies sniff their new home.

I shrugged. After the Harold dinner, Bill couldn't stop talking about when we were going to raise another turkey. He loved the taste of the heritage turkey and the rich stock I made from the carcass.

But I, being on the production end of things, was a little less smitten with raising turkeys. Harold and Maude had been expensive. Bill hadn't seen the receipts from the farm store where I bought the meat-bird feed. The feed had been like a new dress on *I Love Lucy*—an extravagance that would've set Ricky to yelling and swearing. The rabbits, on the other hand, were going to be relatively cheap.

"Breed them and eat them, I guess," I said.

Bill made a face. Every Thanksgiving for the past thirty-six years of his life, he had eaten a turkey. He hadn't grown up eating rabbits.

But I had. And compared to Harold, the rabbits were going to be very easy. There were no rescue trips over to the neighbors, no turkeys chortling from rooftops. The bunnies hunkered in their pen, meek and skittish.

I poured the alfalfa-pellet feed into a dish and filled the waterers Nico had given me—plastic containers embossed with the word "Lix-it." For bedding, I tossed in some fresh straw, which they immediately pounced on and began to nibble. They seemed quite content.

The next day I rode my bike to the downtown Oakland Public Library. It sits on a busy corner in a nicer part of Oakland, the part with coffee shops and baby-clothing stores and Lake Merritt, a jogger's paradise and a bird-watcher's dream. The library was built in the 1950s and has a modernist feel to it. I walked past the homeless people using the computers and climbed the marble stairs to the third floor, past some framed black-and-white photos of historic Oakland. I paused at one, a shot of five men holding loaves of bread, one with a glass of wine in his hand. THE SCAVENGERS, a sign on their truck read. These were the famous Italian garbage collectors of Oakland; someone had told me about them at a dinner party. In the 1920s, they scooted around Oakland collecting garbage. They recycled everything—food scraps went to hog-raising operations, paper and metal were separated and reused. Even clothing was cut up and used for scraps. Now, that was American thrift at its finest. Some of them probably raised rabbits, even. I leaned in for a closer look. Their bread looked pretty good.

Down the hall, I entered the musty History Room, a cramped space filled with books that you can't check out.

"A little young for that, aren't you?" the librarian, a pretty blond lady, asked when I handed her the slip of paper with the call number for the *Whole Earth Catalog*.

"Yeah, my mom always talks about it," I said. The periodical had been one of her main resources for country living on the farm in Idaho.

My parents weren't the only ones in their day to move to rural areas and try to live off the land. By some estimates more than one million young people in the 1970s moved out of cities and tried their hand at farming. The *Whole Earth Catalog* had been like the Internet for this generation of wannabe farmers.

After a few minutes in a back room, the librarian emerged and hefted a massive tome into my hands. It was oversized and featured a photo of the moon. On the back cover the words "Stay Hungry. Stay Foolish" were embossed across a photo of a country road running next to some train tracks.

I sat at a solid wooden table and flipped through the catalog's enormous pages. Reviews of Gregory Bateson and E. F. Schumacher and Thoreau's *Walden* ("The prime document of America's 3rd revolution, now in progress"). An ad for a book called *Should Trees Have Standing: Toward Legal Rights for Natural Objects,* by Christopher Stone. Whole sections devoted to windmills, composting toilets, and other tools of the hippie revolution.

On page 66, I reached the rabbit section. Along with reviews of the Farmers Bulletin No. 2131, *Raising Rabbits,* it included something called the "Rambling Rabbit Rap," written by Gurney Norman: "Raising rabbits is play, it's fun, a hobby. But it can also be work, good, productive work of the kind that contributes to health and vigor by getting good home-grown food on the table."

That had been one of my parents' main goals: to be self-sufficient, to raise their own meat and milk, to build their own house. This desire was a cultural virus, part of the first ecological movement in the United States.

I flipped through the *Whole Earth Catalog* with growing interest. One female rabbit, I read, could have up to thirty offspring in a year. They enjoy shady, cool conditions. Don't feed them cabbage. Building rabbit housing is fun and easy.

The History Room, full of coughing scholars turning dusty pages, suddenly became a vibrating, living place. These old words weren't just memories; they were still useful. I took down notes, pledged to Google the Farmers Bulletin No. 2131, and became increasingly convinced that rabbits might just be the perfect farm animal.

CHAPTER THIRTEEN

❀

When the weather warmed, I donned my bee veil, set fire to some burlap in my smoker, and went out to the deck to perform a hive inspection. I had noticed there weren't many bees flitting in and out of the bee boxes, but this was normal: when the weather is cold and wet, bees usually don't venture out much.

The hive inspection is a springtime ritual for beekeepers. Mostly we want to see how our queen is performing—by early spring, she should be laying a circular pattern of eggs in the brood chamber. The bees' larder of honey and pollen is checked, too. If there isn't enough, they might need a sugar-water supplement to pull them through. This supplement is delivered through a mason jar with a lid riddled with tiny holes; the jar is filled with equal parts sugar and water and inverted to cover a hole on the top of the hive.

I held the smoker near the entrance of the box and squeezed the bellows to let out a few puffs. The smoke has a calming effect on the bees. It is thought that smoke worries them into eating honey (the hive is on fire!), which distends their bellies, which makes stinging difficult. But I've noticed that the smoke mellows them instantly, so it's hard to know exactly.

I pried open the lid with my hive tool, which looks like a thin metal spatula. The lid creaked open, well coated with propolis to keep drafts out. Inside, there should have been bees—moving across the frames, doing clean-up work, making new honeycomb, trying to sting me, an invader. There were no bees. I heard a faint echo of a buzz at the very bottom of the box.

Frantic and feeling sick to my stomach, I pulled off the uninhabited top super and set it on the floor of the deck. Then I pried up the middle frame from the brood box. The brood box is the bottom-most container and is deeper than the honey super boxes. It gives the queen a larger area in which to lay eggs, of which she can sometimes deposit twelve hundred a day. Down at the bottom of the dark chamber was a fist-sized cluster of bees, huddled together.

Smoke curled into their chambers like a fog. I wrenched off my veil and pulled off my gloves. I prodded the bees with the tip of my hive tool. I wanted the cluster to come to life, to attack me. They were entirely docile—nothing to defend.

I started to look for clues to what had doomed the colony. I tipped up each of the brood frames. The top edges were decorated with concentric circles of mustard yellow, almost pure white, and bright orange—pollen, like pop art. A few of the frames were lined with dark-colored capped honey. But there was no sign of brood, the white larvae of the honeybee; none of the honeycomb contained the chubby yellow cells that indicate pupating larvae. Alas, the queen must have died. I put the frames back into the bottom box over the cluster of survivors. They wouldn't make it.

Just as I snuffed out the smoker, ten black-and-white cruisers careened up MLK. For an instant, I thought they were coming for me. When I lived in Seattle, I had a fairly sizeable marijuana-growing operation in the attic. It had made me very paranoid. So seeing the police, even all these years later, caused my heart to race. But what would they get me for this time? Killing a turkey without a license? Too many chickens? The death of my bee colony?

They squealed past the 2-8 and stopped in front of an anonymous warehouse kitty-corner from the garden. Some of the cars were emblazoned with CANINE UNIT. I had always wondered about that warehouse. No one seemed to live there, but at night a guy with a brand-new SUV would sometimes idle in front of the building. Probably a grow operation.

Two plainclothes police pounded on the building's metal roll door. The other officers crouched behind their cars and slowly moved in. The police inched closer. The door came up cautiously—and then, from the deck, I noticed the peach blossoms.

They were frilly and deep pink. The peach trees, a gift from the monks, grew in the parking strip between my garden and the just-raided pot ware-house. And now they were in bloom. Bobby had helped me plant them, say-ing, "We're going to have us some peaches!"

Then I noticed the other trees. A weeping Santa Rosa plum, branches like dreadlocks decorated with white blossoms. The three-way-grafted apple, with its girlish pink and white blooms, each promising a fruit, each branch a different variety. Even the eucalyptus across the street, throwing shadows on the police, was adorned with thousands of filamenty flowers.

I looked down at the vacant beehive, sprawled apart, empty, there on my deck. All those blooms but no honeybees. I put the empty hive back together again. The boxes were getting old, I noticed, paint chipping off. They were looking as tattered as some of the houses in GhostTown.

Across the street, there appeared to be quite a few plants behind the doors of the warehouse. The police were prodding them. Growing pot for medicinal use is legal in California. There are some rules for indoor opera-tions, though, including only ninety-nine plants per building. Maybe this warehouse had gotten too ambitious.

It was funny to see so much green behind a metal roll door, framed by so much concrete and city. My squat garden was the same way. Incongruous. The police started to load up the plants in the trunks of their cars, to evacu-ate the building of its life.

I knew there were other creatures—native bees, flies, ants even—that would pollinate my crops. But the death of a hive left me feeling sour and alone.

When a beekeeper dies, someone has to tell the bees. I learned that at a beekeeping museum in Slovenia years ago. I had been in Europe for my sister Riana's wedding to Benji, her French husband. After the wedding, my mom and I had traveled through the former Yugoslavia. One day, exhausted from sightseeing, we stopped into the beekeeping museum near the town of Bled. We rested our feet and watched a movie in the back of the museum.

The film was shot in golden light. A grandfather showed a lederhosen-wearing boy how to be a beekeeper. It was instructional—how to install a hive, harvest honey, and winterize the bees in snowy Slovenia. Near the end of the film, my mom and I were startled when the grandpa character died.

In the last scene, the boy hunkered near the hive, his lips moving in a whisper. His grandpa had told him that the bees would need to know of his death. The whisperer would feel the heat of the hive, generated by so many thousands of bees. He would smell the wax and propolis. Hear the noise of the bees, as if they were wailing, too. I could see how this act would be consoling in the face of death.

When the lights came up, my mom and I cried a little in each other's arms (we tend to get a bit emotional about things like this) and then proceeded, in the true American response to death, to buy up most of the museum's bee-related merchandise. Posters, honey wine, and folksy hand-painted hive panels.

The movie my mom and I had watched in Slovenia didn't address the tragedy of the death of a hive, though. Maybe the bees never died in Slovenia. I left the empty hive on the deck, another failure. The smell of smoke clung to my clothes the rest of the day.

CHAPTER FOURTEEN

While I grappled with the death of my hive, my sister, Riana, welcomed a new life—she had given birth to a healthy baby girl. From France she had been sending me adorable photos of her little munchkin, Amaya Madeline, and I had to go see her. I saved my pennies for the flight, and Bill agreed to hold things down at the farm.

Five years earlier, my sister had abandoned America. She had been living a life of excitement and excess in Los Angeles, full of parties, Botox, and extreme waxing. My mom had worried about Riana, who, in the ultimate reaction against hippie values, had become a materialist working for a high-end department store and driving an SUV. But then she met Benji in, of all places, the Paris hotel in Las Vegas. I found it amusing that I rediscovered bacon in Las Vegas, while my sister found true love. "I knew the minute we kissed on the dance floor, sober," Riana said, "that he would be my husband." Now she and Benji lived a quiet life in a seaside village in the south of France near the Spanish border. He worked as a math teacher, and Riana became a travel writer.

In March, after a fourteen-hour flight from San Francisco to Barcelona, I took a train from Spain to Narbonne, where Benji picked me up from the station. Like any immigrant, Riana missed a few key things from her native country, and so I entered Europe bearing the tastes of home: baking powder and, though they aren't technically American, corn tortillas. The road from Narbonne to St. Pierre, where my sister lived, passed rolling hills of wild thyme, oak woodlands interspersed with grapevines. I ignored the beautiful scenery—I couldn't wait to hold that baby.

At the apartment, which had been the summer home of Benji's parents,

I clambered up the stairs. My niece, Amaya, was exquisite in her purple onesie. Large eyes, dark hair, olive skin. She looked just like Benji. And yet there was a hint of Elvis to her—a little pompadour, a snarl to her lips. Yes, she was American, too. I buried my nose in her baby neck and greedily smelled her baby essence. Then I passed out on the couch from exhaustion and jet lag.

The next morning—after I had settled into their place, learned how to work the espresso machine, and ventured out briefly to explore the sleepy seaside village—Riana pried Amaya out of my arms and sent me and Benji to the market in Narbonne. Our mom was arriving that afternoon, and Riana wanted to make her a special dinner.

Les Halles in Narbonne reminded me that I wasn't in Oakland anymore. There were stacks of sardines, still shiny from the sea. There were the *boucheries chevaline,* the horse butchers, selling blood-red cubes of horseflesh. There were reasonably priced goat cheeses, solid blocks of tomme, and ripe Époisses. The chickens had their heads and feet still attached. The French housewives sashayed through the covered market stalls and stopped to examine the scales on the legs of the chickens before buying.

"To make sure it is fresh," Benji explained, his French lips lingering irresistibly on the "sh" sound.

We were an unlikely pair—Benji suave, calm; me wild-eyed, loud, and slobbering. We paused at the rabbit stall. Skinned bunnies lined a glass case, their heads still attached. So that's how they look underneath their fur, I thought. They were pure muscle, with no fat. Their back legs looked like skinned chicken thighs.

The rabbit farmers had mounted photos of their idyllic south-of-France farm on the wall. Their place looked nice—rolling hills, a stone farmhouse. The lady behind the counter smiled and greeted me. "Bonjour!" I said, then worried she might say something else. I couldn't even say I didn't speak French in French. It was a comfort to have Benji at my side.

"Benji, will you ask them how they kill the rabbits?" I said, and nudged him.

Benji sighed and reluctantly said something in French to the woman. She looked a little surprised. Benji laughed nervously and said something else and pointed at me. The woman cleared her throat and looked at me while she made some brisk sawing motions with her hand while explaining.

"She said they make a slit in their throat," Benji reported.

"They don't bash them in the head?" I asked. "Or break their necks?" Benji asked in French. The woman looked mortified. Who was this barbarian?

"No." Benji has these big brown eyes, and they were cast downward in shame.

"Sorry, Benji," I said.

I was sorry to embarrass Benji, but I had to figure it out for my own project at home. The *Whole Earth Catalog* had been silent about how to actually kill a rabbit.

The French rabbit lady nodded her head when we pointed at a plump bunny in the case. She took out an enormous pair of scissors—I mean enormous: the blade was almost two feet long—and cut our rabbit up into pieces like a chicken.

At my sister's insistence, we brought the rabbit's head home. Though many French people do eat the head—the cheeks and the brain are particularly toothsome, according to Benji's grandmother—this one was for Lucky the cat.

Back at the apartment, Riana announced that Mom had arrived while we were shopping and was taking a nap. While Benji unloaded the groceries, I held and bounced the Buddha-like Amaya on my knee. I couldn't get over the fact that my sister had become a mom. I wondered if having a baby around had any similarities to raising farm animals.

Riana dredged the pieces of rabbit in flour, thyme, piri-piri. Her blond hair up in a bun, she stood at the gas stove wearing an apron and fried the rabbit in hot oil. Growing up, we had always been mistaken for twins, even though we are separated by eighteen months—we both have angular faces with prominent chins and tend toward the tall and leggy side of things. My sister, who has always been a confident person, both in cooking and in traveling, now had a slightly different aura. She just seemed more authoritative than usual. In her time in France, I could see that she had really blossomed. And her cooking had clearly reached another level of deliciousness.

Riana and I had always been into cooking. We could make a marinara and a béchamel sauce before we hit puberty. We learned to cook because we had to. When I was ten and Riana twelve, our mom was diagnosed with multiple sclerosis. One morning, she went blind in her left eye and had to wear a patch. She had trouble walking. And since she was a single parent with a full-time job and a sickness that left her exhausted after a day of work, my sister and I learned to make dinner.

My parents had divorced in 1977. My mom then took us to Shelton, Washington, a rainy logging town near Seattle, where she got a job as a schoolteacher. Times were tough—teaching didn't pay well, and my dad didn't pay child support and rarely saw us. I remember overhearing my mom talking with one of her friends. "You're like a mother wolf, taking care of your babies," her friend had whispered. My mom protected us and wanted us to thrive.

The image of my mom as a wolf stuck with me, because my mom brought home all manner of free food for us. She worked at a school on the Skokomish Indian reservation, so her students often gave her whole salmon caught in the Skok River and sacks of oysters from Hood Canal. She and friends would go out on chanterelle-mushroom-picking expeditions. My mom also grew—and had Riana and me tend—a large kitchen garden next to our house. It was a small reminder of her ranch days, except instead of fields of corn and tomatoes, it was just a few rows.

The door to the guest bedroom opened. My mom, jet-lagged, staggered out. She sniffed the air with her long nose, which I inherited.

"Smells like rabbit," she said. We hugged, and then Mom sat down at one of the kitchen stools to watch Riana cook. "I remember when I would take you two girls out to the rabbit hutches . . . ," she began, fingering her long, dangly turquoise earrings. Despite her jet lag, she was awake enough to recount another one of her farm stories.

Riana glanced at me, and we simultaneously rolled our eyes.

We've heard all the Idaho farm stories so many times that if Mom starts one, Riana and I can recite it verbatim. The time Zachary the dog killed the chickens and Dad had to shoot him. How we would watch my mom milking the cow, waiting with bottles in hand for our milk fix. And, yes, the rabbit butchering, when Mom, wearing a down coat that she had stuffed herself with goose feathers from the farm, would give us a tour of the inside of a rabbit. "These are the small intestines; this is the heart," she would instruct as she'd point with her knife tip, the rabbit, tied to a tree branch and flayed open, steaming in the cold air.

There's also a piece of photographic evidence for this ranch tale: Riana and I standing in a snowy glade with bad haircuts, Riana holding a big white and black rabbit in her arms. These animals were not pets.

But that was a long time ago. Along with most of the other back-to-the-landers, my mom had realized that the remaking of our entire American society might not be possible in her lifetime. That spinning wool or churning butter might be fun for a while, but eventually the conveniences of modern life—grocery stores, clothes driers—seemed pretty wonderful. The possibilities for mockery, in hindsight, are endless. The back-to-the-land movement's failure, as inevitable as the collapse of every other utopia, became a buffet of schadenfreude at which even I had occasionally feasted.

But now that I was farming, I knew it was hard work and that plans never went the way you thought they would. After the Maude tragedy and the watermelon debacle, I would never laugh at my parents' hapless experiment again. I'm sure my mom had many a run-in with an opossum—and that shit is not funny.

Most of my memories of the farm disappeared in the 1980s, replaced by neon-hued socks and crimping irons. But our mom kept the idea alive with her endless retelling of farm stories.

Although Riana and I give her grief for it, I could see why she did it. Her time on the farm had been filled with defining moments: the first beam raised in their house, her first homemade cheese, her first baby. It was an era when creatures had become characters in the fabric of her life, when the apple harvest meant there would be fruit throughout the winter, when a rabbit raised and slaughtered behind the house meant both a biology lesson and a tasty dinner. There was a lot of room for nostalgia. It was also a time when she was young and healthy and could do anything. And so Riana and

I let her tell her stories, out of respect and sometimes curiosity, and tried to imagine what she had been like then.

In honor of Mother, Riana was making *civet de lapin,* rabbit in blood sauce, a step up from how it was usually prepared on the ranch in Idaho, fried like chicken. Riana put the browned rabbit into a tagine, a ceramic cooking vessel. The still-raw liver went on top, and a bottle of wine was poured over the whole thing. This all was covered with the smokestack lid of the tagine and whisked into the oven.

We sipped the local rosé and watched the sun dip into the Mediterranean. My sister and I dutifully listened to my mom tell the rabbits-on-the-ranch story again, happy to be together, making new memories in France. I half-heartedly wished that my dad could have been there, too. He spoke perfect French—he had studied for a year in Grenoble when he was a young man. After he and my mom got together, they traveled through France. Not far from my sister and Benji's home, my parents had picked grapes as hippie gypsies. My mom loved to tell the story about how the other pickers would call her the Snail. She was slow because she was pregnant with my sister, and she had to periodically stop her work and quietly vomit into the grapevines.

My sister was born in Idaho but had, thirty-five years later, found her way—all this way—back to where she had been conceived. It is my mom's—our family's—most amazing story.

Later that night, Riana was up with the baby. Since I was sleeping on the couch next to Amaya's crib, I was up, too.

"How did Mom do this?" Riana said, looking down at Amaya nursing. While Riana couldn't relate through farming, motherhood had made her see my mom in a different light.

"Dude," I said, "they didn't even have electricity."

"And they—we—lived in the trailer while they built the house," Riana whispered. "That tiny trailer," she said, and wiped Amaya's chin. "I can barely cook dinner with a baby, much less *build a house.*"

"All those animals," I added. Our minds were boggled at our parents' moxie.

That night, lying there on the couch, I thought about my life in Oakland

and its general trajectory. My parents had, by my age, built a house from scratch, had two children, and fed themselves from their land. My sister had, in the past five years, gotten married, given birth to a beautiful child, and learned to speak fluent French and cook flawless French food. I, meanwhile, had some raggedy chickens, some borrowed rabbits, and a dead beehive. On land that could be bulldozed at any moment. My peers were homeless people and freaks.

In France, I noticed that I had even come to pick up some of the patois of the rough-and-tumble streets of Oakland. At dinner, I found myself saying "How you?" and "Hella cool." My clothes were stained and starting to disintegrate—part and parcel, I suppose, of being an urban farmer.

However, even that identity, viewed from a distance, was starting to seem rather . . . thin. When I explained to my sister and mom that I was an urban farmer now, I could see that they had concerns about that self-definition. Because whom was I really feeding? Yes, I had successfully raised a perfect heritage-breed turkey, and it had been delicious. But was there any evidence that I could actually feed myself on a day-to-day basis? I was young and healthy, in my prime, I could do anything, and I was ready for a challenge.

Around 2 a.m., a reckless thought about self-sufficiency came into my head. It niggled at my brain while I tossed, wide awake, on the couch. It made me do some math involving rabbit-breeding cycles. In the morning, over the first of many cafés noir, the idea hatched: for the month of July, when the first of my so-far unborn rabbits would be ready to harvest, I would feed myself exclusively off my urban farm.

"Hey, Riana, can I get that rabbit recipe?" I asked, rocking Amaya in my arms.

✷

CHAPTER FIFTEEN

I arrived home from France on a Wednesday night in April. Bill picked me up in our jalopy. Though it was late, I was wide awake. I also had a smuggler's high from successfully getting contraband stinky cheese and cured duck breast from Les Halles past customs, wrapped in my dirty underwear and socks. I waved a Ste. Maure goat cheese in Bill's face as he drove.

"Vella, I got bad news," he said.

My stomach dropped. I immediately thought of the rabbits, the chickens, our cat. Dead. Or that Jack Chan had reared his real-estate-developing head again.

"What?"

"Lana's moving away."

"Oh, no."

Bill took the overland route instead of the highway. As we cruised up MLK, I reacquainted myself with the sprawling garbage, the guys pulling shopping carts, the drug dealers on the corners. I had only been gone for ten days, but GhostTown looked grittier than I remembered. I wondered what Benji would think of this place.

When we pulled up to our house, I suddenly had a fear: Was my diamond in the rough actually a cubic zirconia in a pile of shit? Had I been deluding myself? I pushed past the gate to the garden. The air that greeted me smelled fresh and clean. Even though it was dark, I knelt to examine the lettuces growing in the raised beds; they were sturdy and vibrant. I sniffed at the sweet peas that sprawled up a trellis. The garlic shoots, I was pleased to see, had grown a few inches. Yes, yes, this was a worthwhile project.

While Bill carried my backpack upstairs, I went around to the chicken house with his flashlight. "Hello, hello," I said at the door, preparing them for my intrusion. They clucked and made a high-pitched trilling noise. Chickens are immobilized by the dark. I shined the light across the perch, catching glints of their feathers, a chicken eye, the cocked comb of another—all huddled together. They were fine. I shined the light down to the rabbits. They ran around in circles, biting each other. I squatted down closer and saw that they were—yes, that's what they're doing—humping. Does on does, a doe humping Simon the buck. I would have to separate them tomorrow. The humping was good news, though. It meant they were ready to start breeding.

After checking on the animals and reassuring myself that the farm was worthwhile, I went over to Lana's. A sign posted on the metal door of the warehouse said the speakeasy had closed. Years ago, Lana had given me the key with permission to enter at will, so I let myself in.

Inside, Lana and her sister were sitting near the faux fireplace, the clutter of fifteen years billowing around them.

"Everyone just gets drunk," Lana said when I asked why she was calling it quits.

"No one performs anymore," I agreed. I had stopped going months and months ago.

We heard some frantic knocking at the door. Lana ignored it.

"Let's burn the couch," Lana's sister said to cheer her up. "Under the overpass." Lana shook her head no.

Lana told me she was moving to Mexico. I nodded—I had seen it coming. She had recently lost Maya and had been devastated. I helped bury the guinea pig in the garden. We interred her next to Maude and the duck and goose. Lana placed a large pair of praying hands on the grave to mark it. While we buried the little brown and black guinea pig, I couldn't help but think that people in South America eat guinea pigs. I was terrible.

Sitting at the bar talking with Lana and her sister, I had a horrible thought: Were my animal-killing ways causing her to move away?

"I'm sorry about killing Harold," I told her. Not that I had killed him, but that she was upset by this act.

"It's better than most meat eaters," said Lana's sister. "At least you faced it."

Lana shook her head. But I knew I had bummed her out. She was like a child in her love of animals. The day after I killed Harold, Joel called to say Jackson woke up in the morning, pulled a turkey feather from underneath his pillow, and cried, "I miss Harold!" Jackson pledged to never eat an animal that he had known personally. Joel and I sighed. Another plan had backfired—did this mean he would insist on factory-farmed meat exclusively? I had hoped, in the back of my mind, that I would become for Jackson like one of my mom's friends whom I fondly remembered from my childhood. Now I was afraid his only memory of me would be a ghoulish, frightening one.

Lana and I looked through photos, and I helped her pack. She ordered a pizza, and Oscar barked at us until we fed him a slice. She found one picture of us standing in the clearing of the lot before any of the beds or plants had gone in. Because of the angle of the photo, we looked like homesteaders on the prairie. The grass and weeds were a tawny gold.

I didn't know how to thank her. She was a big reason we came to live on 28th Street. She had been directly and indirectly responsible for so much of my happiness.

"Lana, I'm going to miss you," I said, unable to think of anything better.

But as I walked back to my apartment I knew that with Lana gone, as much as I would miss her, my experiment in self-sufficiency—in proving to myself that I was a real farmer capable of feeding myself—was going to be so much easier.

Rising at dawn because of jet lag the next morning, I went out to our seventy-five-square-foot deck, where the defunct beehive still sat, and created a rabbitry: a series of tunnels and boxes, hutches and cages. I threw hay and tossed sawdust onto the deck floor, which was made of rough roofing material. To add a festive air, I hung a clattering bamboo and coconut-shell wind chime over the whole thing.

Then I got the rabbits.

Adult rabbits, I had read in the *Whole Earth Catalog,* need to live in their own private quarters. They are considered adults when they start humping each other. If I didn't separate my rabbits, Simon would relentlessly try to breed with the females, and the females might kill each other's babies, maybe each other.

At the chicken-run door, I held out a stalk of celery. Simon hopped over. His nose was just like one in a children's tale—remarkably dislocated from his body, and fuzzy. Like cats, rabbits have a flabby layer of skin along their necks and backs that makes a great place to hold on to. I just had to get close enough for a grab. I petted Simon, but he seemed uneasy. Tentatively, he pulled the celery out of my hand. Then I collared him.

Instead of running, Simon tensed up every muscle in his body so I couldn't get a handle. Buying that critical second, he heaved to the far side of the hutch. The females cowered in the corner.

I had to go in after him. By turning my shoulders, I crammed my five-foot-eight frame through the small doorway of the rabbit run so half my body was inside the cage, half out, and managed to grab Simon. As I shimmied out, for a second I had the irrational fear that I would be stuck inside this cage, my legs dangling out. The chickens would eventually start pecking me, the ingrates. But luckily, my hips cleared the door with no problem, and Simon and I left the cage together, farmer and bunny.

His legs drew up and his body curved into a C shape. His fur was impossibly soft. Wading my way through the chickens, I cradled Simon close to my body. He, in true rabbit style, tucked his head under my armpit. If he can't see what's happening, nothing bad will happen. Fuzzy logic.

When I opened the door to his very own cage (feathered with timothy hay, straw, wood chips, one of my old wool sweaters, and his personal water bottle), he arched his back and pushed both his hind legs off my body to leap into his new home. His feet have claws—remember those '80s rabbit's-foot keychains?—and I winced as they ripped into soft flesh.

"Hi, Novella!" Lana yelled from across the street. I waved back, standing in front of the rabbit cage so she wouldn't see the newest meat on the farm. She had a box in her arms and added it to a growing pile on the sidewalk, then disappeared back into her warehouse.

I looked down at my Simon-inflicted wound. Two parallel scratches, four inches long, puckered my forearm. A bit of blood oozed out. A man in a truck with an enormous MICHOACÁN bumper sticker pulled up in front of Lana's.

I went downstairs to say goodbye. Lana seemed calm, determined even. Her hazel eyes were a bit red when we hugged. She eyed the scratch on my arm but didn't say anything. I made plans to pour some hydrogen peroxide on my wound.

Lana gave us some stuff from her house: a giant puppet hand, a cracked salad bowl from Italy that had been glued back together, some espresso cups. The guy with the truck attached her bike to the truck bed with a bungee cord, then Lana climbed into the vehicle with Oscar the dog and was gone forever. I saw her in profile as she left, looking forward, her chin jutting out a bit. Oscar stuck his head out the window.

One by one I relocated the female rabbits. They each got their own box (to hide in), water bottle, and food dish. I put the cages close together so the rabbits could smell one another. The deck was utterly transformed. The straw on the floor glowed gold. The rabbits scurried around in their private cages, smearing their noses against the new surfaces. Simon thoughtfully chewed a piece of celery clutched between his paws. My deck looked like a third world country. And I liked it.

Downstairs, while I was watering the garden, I heard a commotion on the street. It was Bobby going through the boxes of stuff Lana had left in front of her house. It looked like he was rearranging his living situation, moving the television over to a table he had set up directly in front of Lana's former gate.

We Americans relocate with impunity, most of us on a regular basis. I thought about Benji, my sister's husband. He still lived in the town where he was born. His great-great-grandfather lived there. But in the States, an idea strikes and we're gone.

When Lana left, it was as if one of the biggest trees in the rainforest canopy had fallen. She had lived here, in one place, for seventeen years, a record for our block. But this is city life—when someone leaves, another rushes in to take her place. In the vacuum of Lana's absence, Bobby took over. Within days, the end of the 2-8 resembled a bingo hall, with all manner of tables and chairs set up. His ever-growing collection of tires and shopping carts sprawled across the dead-end street. What had once been Lana's sidewalk had now become Bobby's domain.

❋

CHAPTER SIXTEEN

A couple of weeks after Lana left and Bobby unfurled like an enthusias-tic kudzu vine, I built a rabbit love nest. It was April, so the rains had stopped, but the grass was springtime lush. If I was to be self-sustaining on the urban farm starting in less than three months, I needed my breeding-stock rabbits to actually breed.

In the squat lot, I erected an enclosure of chicken wire in a half-sun, half-shade spot, then kicked in a red ball as an icebreaker. I placed Simon in the love cage with one of the speckled brown and white does. At first they were shy. Though they had grown up together, they were now living in separate hutches. They sniffed each other, explored a little, and tasted the different varietals of grass and clover within the bounds of the enclosure. I noticed they especially liked Lana's weed, the *Malva parviflora*. To achieve the best purchase on the spiky weed, they had to clamber over each other. Then, I hoped, one thing would lead to another.

Rabbits, I had come to realize after some reading, had provided meat for people hell-bent on survival farming way before the hippie back-to-the-land days. During World War II, "thousands of resourceful Americans raised rabbits in backyards to put meat on the table when ration stamps were not sufficient to do the job," wrote *Raising Rabbits Successfully* author Bob Ben-nett. "In cities and towns, in the best of neighborhoods, rabbits were housed in wood and chicken wire hutches, busily putting meat on their owners' tables." In the best of neighborhoods.

As I weeded and planted near the new lovers, I sensed that they might feel exposed away from their safe pens, suddenly out in the garden with open sky and the whizzing sounds of the highway. I added a bucket to the love cage.

Just then, the woman Hillbilly walked by with her Chihuahua and beckoned me over to the sidewalk. After Lana had left, I finally found out the Hillbillies' real names: Peggy and Joe.

"Are you still doing the community garden?" Peggy asked.

"Yes," I said. I tucked my dirty gloves into my back pocket. "Do you want to have a plot?" I asked. So far Mr. Nguyen had been the only person to take over and dutifully tend one of the raised beds.

"Oh, no, no. We don't know how to do that!" She laughed.

The dog took a tiny poo on the parking strip. Then Peggy rustled around in her coat pocket. "But we'd like to make a donation," she said.

I started to protest—this operation was essentially free, except for my time—until I saw the seed packets.

Tomatoes, brussels sprouts, cucumbers. I noted that they were hybrid seeds, Burpee.

"Can you grow these?" she asked.

I nodded and took them from her hands, noting that "$1.49" was written on the corner of each packet. I realized that this was her way of putting in her order for the summer harvest.

"Sure," I said. I had become a farmer for hire.

On a break, I went upstairs and tossed the Hillbillies' seeds in a box with all the others (expired, free, inappropriate for our area) I planned to discard by randomly throwing them onto vacant properties around our neighborhood. If some of them made it—great. I didn't want to be a snob, but there's something unsavory about hybrid seeds. Many of those sold by seed companies are F1 hybrids. This means that the seed is the offspring of two inbred parent plants. Inbreeding tends to weaken seeds, but scientists figured out a long time ago that if you breed two inbreeds, you will get a plant that exhibits "hybrid vigor"—it will grow really fast and strong and uniformly. However, you can't save the seeds from such plants, because their offspring, referred to as F2, are usually weak and not uniform. To get the strong F1 seeds, you need a professional to breed one.

Instead of the Burpee tomato seeds, I'd plant some Bill had saved from a Brandywine that had performed particularly well in last year's garden.

He harvested the biggest tomato on the vine, squeezed out the seeds, then set them on the windowsill to rot. After a few days, they had grown mold, which ate away the protective seed coat and ensured better germination. Then Bill had washed them off and squirreled them away for the next year.

Heirloom varietals come with cool stories. The Brandywine seeds sold by heirloom seed companies today are descendants of those an octogenarian seed saver in Ohio named Ben Quisenberry got in 1980 from a Mrs. Dorris Sudduth Hill, who said they had been in her family since 1900. To muddy the waters a bit, farmers and seed savers over the years have created other strains, like the heart-shaped Brandywine, the yellow Brandywine, and the cherry Brandywine. There's the Sudduth strain and a pink strain, though the latter is said to be an inferior producer. As confusing as it all can be, the Brandywine is a living example of how messy, how fertile, how diverse heirloom seeds can be.

Bill and I got our first Brandywine seeds from a seed swap in Berkeley our first year here. They grew up to have leaves as big as those on potato plants and large, slightly misshapen red fruit. It was meaty and juicy, only slightly tart. That same year, just to see what would happen, we also grew, from saved seeds, some Sungold hybrid cherry tomatoes. But instead of producing delicious orange fruits that taste like pineapple, they yielded strangely small, bitter red fruits. If I wanted a true Sungold, then, I'd have to shell out some money to a seed company. It's not like that with heirlooms, which breed true and can be passed from farmer to farmer, generation to generation, with no middleman. Heirlooms are different from hybrids, too, in that they can adapt to local conditions. That's why saving seeds from a plant and planting them in the same soil and climate from which they grew will make an even stronger plant.

I checked the cupboard where I kept the seeds and found Bill's Brandywine. I had also recently gotten in an order from Seed Savers Exchange, an heirloom-seed company that capitalizes on romantic, old-timey vegetable stories. I usually bought my seeds at cost from the nursery where I worked or through a seed swap at the Ecology Center in Berkeley. But the temptation of a seed catalog—vegetable porn, really—always overwhelmed me, and I usually ordered something truly exciting.

Peggy wanted cucumbers? How about the Boothby's Blonde cucumber—pale with black spines—which has been grown in Maine for generations? As for brussels sprouts, I had never had much luck with them, but I would research some heritage varietals, and Peggy and Joe would love them.

I poured myself a glass of water and walked over to the window. I looked out into the lot to check on the bunnies and saw that they had made it past introductions. Simon was furiously humping the speckled rabbit's head. His cotton-tailed hindquarters pumped for a minute before he collapsed backward in an exhausted, furry heap. I took a sip of water. Simon might not cut the mustard for my rabbit-breeding program.

Early the next morning, a cop car, a city car, and a tow truck arrived on 28th Street. A dump truck idled nearby. Bobby was pulled out of his car by the police officers. And then he watched them take his world.

The umbrella and table, the shopping carts filled with metal, car parts—burly men wearing green city-worker coveralls tossed these items into the dump truck. They threw everything else—the television, the microwave, soggy stuffed animals, pillows, tennis rackets—into the four broken cars Bobby had been sleeping in. Then they towed the cars away. A man with a clipboard took notes of the proceedings. A cop stood next to Bobby, who kept lunging at choice items, trying to save them. "Don't touch anything," he said, holding up his arm.

Then Bobby just slipped away through the schoolyard. I watched, like a coward, from our living room window.

I wondered who had finally called the city on Bobby. It might have been the owner of Lana's warehouse. Bobby's spread wasn't very attractive to new renters.

It's true that Bobby had been a nuisance. Parking had become impossible on the 2-8 with all his cars. The smell of urine was unmistakable near his camp. But he was a lovable nuisance. The city didn't know, for instance, that Bobby kept the street swept of glass and helped the garbage collectors loft the heavy bins into the trucks every week, that he helped us push our cars and generally monitored the street.

The city vehicles finally trundled away, leaving an empty, stained street.

Bill hugged me that night when I told him about what they had done to Bobby. To cheer me up, he told me about a new discovery he had made in our neighborhood. He gave me a handmade sign. CHICKEN WING OR CATFISH DINNER, it read, with an address one block from our house. The sign promised sides like potato salad and peach cobbler.

Bill and I went over. The house was cute, in the way "cute" really means: A birdbath. Rosebushes. Pots of flowers surrounded by white pebbles. About eight burly black guys stood outside the gates of the house.

"Is this the . . . ?" I began to stammer.

"Novella!" It was Bobby, who emerged from around the corner. I gave him a hug.

"Are you OK?" I asked.

"Of course," he said, and smiled. Bill went in for a hug, too, but Bobby pushed him away. "I only hug women!"

"Hey, are these dinners good?" I asked him.

"Only the best," Bobby reported.

"Do you want one?" I asked.

"I'll just take a bite," Bobby said.

"For here?" one of the big guys said, overhearing us.

"Yeah."

"Let me go tell Grandma," he said. He ran up the porch steps, then paused and turned back around. "Chicken or fish?"

I looked at Bobby. "Fish?"

"I just want a bite." Bobby grinned.

We went into the garden area and sat at a glass table with a flowery umbrella over it. As we waited for our food we watched pimped-out cars careening down Martin Luther King and a homeless woman grooming herself in the mirror of a parked car. I was pretty sure none of this happens in France.

Bobby had theories about who had called the police. An elderly couple had bought a parcel of land at the end of the street and were planning to build a new house just in front of the area where Bobby was living. Bill and I shook our heads. Development was the bane of our existence.

The dinners came, a glorious Oakland version of Slow Food. The fish was perfectly golden and fresh. "Grandma," it turned out, was a fisherwoman. She came outside to see how we liked her cooking.

"Caught the fish myself," she said proudly. She was about sixty years old and kept her long gray hair in a ponytail. She and her husband, Carlos, would go fishing near South San Francisco, then cook the catfish, bluegills, and striped bass for these neighborhood meals. The fish came with a side of spicy collard greens, a scoop of tasty homemade macaroni and cheese, and a glop of peach cobbler. The meal represented American thrift at its finest.

We gladly paid Grandma's son $10 each for dinner, then Bill, Bobby, and I hunkered over our food in the late spring air. This underground restaurant would never happen in France either.

"What are you going to do?" Bill finally asked Bobby, after eating the last piece of fish.

"You'll see," he said, and gave us his mischievous smile.

At home, Bill and I tried to think of something to do for Bobby. We both were furious that the city wouldn't allow him to live on the street yet didn't try to find him a new place to live either. While we talked about it, Bill said, "Hey, what's that?"

An ancient tow truck backed onto our street. It crawled back, bearing its load—a battered red Taurus. Out sprang Bobby from the passenger seat. The driver eased the car into a parking spot. Bobby unhooked the car from the trailer. He had returned. He grinned to himself. Round one—Bobby.

❊

CHAPTER SEVENTEEN

Two months later, Bill and I found ourselves up to our armpits in the green bins of late-night Chinatown. I wore blue nitrile gloves and was scooping handfuls of bok choy into my bucket. The sound of clicking mah-jongg tiles serenaded us from an open second-story window.

Oakland's Chinatown is less than twenty blocks from our house. It is a second city within a second city. It doesn't have the elaborate temple-style architecture of San Francisco's tourist-attracting enclave. No giant tiled animals announce the borders of the place; no strings of Chinese lanterns hang across the streets.

There are restaurants, but Oakland's Chinatown makes most of its living by selling ingredients: hand-cut noodles, plump dumplings, live seafood, medicinal herbs, and, yes, lots of produce. Giant daikon, a long white type of radish. Pickled turnips sold in rancid-smelling plastic vats. Persimmons. Lychees. Apples, oranges, grapes, bananas.

Amid this abundance, of course, was a lot of waste. Our pilfering from Chinatown's Dumpsters and green bins became a botanical education. There were so many different kinds of greens. Bok choy and Thai basil and Napa cabbage I could identify without trouble. But what was this long, thin-leaved thing? This giant watery-looking leaf? This oniony-smelling grass? Kangkong (water spinach). Choy sum (Chinese cabbage). Nira grass (a kind of chives).

We were trolling the garbage cans for the rabbits. Sure, they dutifully ate their alfalfa pellets, but I had noticed they were happiest when we gave them scraps from our table (apple cores, damaged lettuce leaves) and our garden (aphidy kale leaves, carrot tops). One night, after a Chinatown meal of wonton soup and too many Tsingtaos, Bill and I boldly flipped open a sidewalk green bin. We were just curious. Bill pawed through the bin. It was filled with fresh-looking greens, some newspaper, moldy oranges, more greens.

"Can we eat those?" he asked me, holding up a rumpled cabbage leaf. I imagined actually cooking food out of a Dumpster, grimaced, and shook my head no. Isn't that what Charles Manson and his followers did?

Then we both remembered at the same time: the rabbits. It turned out that they loved the bok choy, radishes, apples, and pears from the garbage cans. Anything the rabbits didn't like (papaya, melons), we gave to the chickens.

Now we were going to Chinatown twice a week to load up on rabbit and chicken food. A few late-night stragglers on Webster Street regarded us—two fairly clean white people wearing headlamps and stuffing leafy greens into buckets—with curiosity. A man approached me and handed me an empty Coca-Cola can. I shook my head at him and resumed sorting through the apples, trying to find unbruised ones for the rabbits. Bill, meanwhile, yelled for me to come over and check out a particularly bountiful bin. We turned the corner and scoped out 10th Street, carrying our swaying buckets like milk maids gone a-milking.

If they could have, I liked to think, the chickens and the rabbits would've hopped or flown over here to 10th and Webster just as most of the restaurants closed and stood there, with furry paws and scaly legs, Dumpster diving. They would hear, as we did, the laughter wafting out of the open windows above us, the smell of cigarette smoke mingling with the odor of fetid fruit.

An ancient Chinese lady wearing elbow-high plastic gloves walked by us. She might have been the same woman Bobby had scared off from our recycling bins. She surveyed what we were doing and raised a thin eyebrow. The Oakland Tribune tower, home to the local newspaper, loomed above us, watching. We snickered, packed our car full of goodies for our animals, and left in a puff of vegetable-oil smoke.

Once home, I delivered their share of the harvest to the rabbits, who by now had greatly increased their numbers. Simon had finally figured out which end to approach and had gotten the hang of sex. And I soon discovered that he wasn't, as I had initially worried, shooting blanks.

A month after her visit to the love cage, the brown and white doe started to act funny. Hoping babies were on the way, I put a nesting box—a double-length milk crate lined with an old wool sweater—in her cage. The doe pulled fur from her underbelly and made a soft nest with it. This behavior is called kindling.

A few days later, Bill pointed to some movement in the airy-hairball nest. When the mama hopped out for a drink of water, we counted the kits. There were eleven rosy pink babies. They looked like worms, eyes closed, squirming blindly around. They grew up to be kittenish, and ranged from pure brown like Simon to pure white to spotted.

One of them didn't make it. The doe abandoned it outside the warm nesting box. Bill discovered the body, cold and dead, on the floor of the deck. It had such an expression of anguish. Its mouth was wide open in a primal scream; its legs were frozen in the motion of kicking. A few white strands of its mother's hair clung to it. Unsure what to do with it, I stuck it in the freezer.

Holding the Chinatown bucket now, I distributed the food to the three females, all of whom had by now kindled and given birth. After filling their feeders with green alfalfa pellets, I handed out apples, one in each cage, then added a heap of bok choy. They fell on the food, seated on their haunches, nibbling, ears quivering.

I tossed Simon less food than the females—he didn't have kits to feed, and I had heard if the males get too fat, they can't breed. Then I went downstairs to unload the rest of the boxes and buckets of veggies from the car.

The chickens were asleep and would get theirs in the morning. The hens loved it when we dumped out a bucket of greens and split open a too-ripe melon for them. They went straight to work, clucking and pecking at the bok choy and melon flesh. They ignored the pale, powdery chicken food.

Their eggs had started to taste richer and the yolks had turned a darker orange since we began our Chinatown runs. Just as exciting: my paychecks from my various jobs were starting to last much longer.

After I closed the trunk of the car, I went out to check on the garden. Saved from the imaginary bulldozers, it seemed more miraculous than ever. It was early June, and the garden beds were stuffed to the point of spewing produce on contact. The peas I had planted in February had set fruit, the fava beans were flashing their green pods, lettuces and chards wallowed in the spring mud, and the new tomato starts were taking hold. It was a lot of food. I had also cracked through concrete to plant two more apple trees and a Bartlett pear, and I had grafted plum scions to the existing plum trees.

I could hear the rabbits upstairs: a glug from the waterer, their nibbling noises. The sounds of satisfaction floated out from the deck and onto the 2-8. There was a new billboard up just down the street. WE BUY HOUSES.

I felt young and healthy, and nostalgic for the present. If urban farming was a competitive sport, I felt as if I was in the zone, at the top of my game, ready for any challenge. If I turned out like my mom, these would be the days that I would recount to my niece—and perhaps my future children— ad nauseam.

CHAPTER EIGHTEEN

✳

A pledge to eat exclusively from a July garden in the Bay Area, I reasoned, is a little like a mute person taking the vow of silence at a Vipasana-meditation retreat. I wasn't worried.

The rules were simple:

1. *Only food from the garden and the farm animals.*
2. *Foraged fruit from neighborhood trees OK.*
3. *No food from Dumpsters (except to feed the animals).*
4. *Items previously grown and preserved allowed.*
5. *Bartering allowed, but only for crops grown by other farmers.*

In mid-June, I told all my friends about my approaching escapade in eating. This, I felt, was critical to its success—and commencement, for that matter. We giggled about my bravado, my moxie, my mad urban-farming skills. While I knew they would be cheering for me, they would also be keeping tabs. Back in March, when I had conceived of this harebrained idea, July had seemed so far away. Now that it was right around the corner, I was starting to think that my experiment in self-sufficiency wouldn't be much fun.

The week before it began, I ate everything in sight. In my excess, I pretended that I was A. J. Liebling in Paris. But I was in America, so I gorged from the buffet of cultures this country hosts. Chinese food, relishing that tarlike sweet-and-sour sauce, the pillowlike dumplings. Sushi. Small chile verde tacos from a roach coach in East Oakland, the perfect blend of pork simmered with green chiles. Falafel, creamy baba ghanoush, tabouli. Every morning I had a huge mug of coffee (sometimes two), brimming with half-and-half. I popped vats of popcorn, scoffed at greens (plenty of time for those later), inhaled chocolate bars, and drank lapsong souchong, a smoky tea whose flavor would be impossible to re-create. This weeklong binge left me a little heavier than my usual fighting weight. In a thrift store, I stood on a scale: 142 pounds.

The evening before June turned into July, I walked out into the garden to survey my future. In *Walden,* Henry David Thoreau wrote, "I was determined to know beans." I too was determined to know beans. I admired their sturdy leaves emerging from the black earth, their raspy stems that wound around whatever kind of pole I could find (currently, a curtain rod), the succulent flowers, and then the emergence of small beans, which could be plucked, blanched, and served plain—because the 100-yard diet didn't allow olive oil or balsamic vinegar.

While Thoreau, no food snob, was happy to cultivate a monotonous crop of beans on his three acres, I was determined to know other vegetables during my month of self-sufficiency on my tenth of an acre. And so I had planted sweet corn, Stowell's Evergreen, which was now about four feet tall and just coming into flower. I hoped some tasty niblets would be mine toward the end of the month. Brandywine tomatoes, too, and the green ones on the vine seemed like a good sign. Prodigious amounts of lettuce, collards, kale, and cabbage had sprouted up all over the garden. I had made a second planting of fava beans. More beans. Henry, did you know these lovelies, brought by Italians to this country? The onions were swelling, as were the beets. Potato vines were peeking out from under a mat of straw. The squash plants had a few young fruits, as did the cucumbers. Herbs like marjoram, oregano, rosemary, and thyme were flourishing.

The domesticated-animal kingdom was a realm in which Thoreau never dabbled. He scoffed that farms are "a great grease spot, redolent of buttermilk!" In my beloved grease spot, one of the chickens was laying an egg in what she thought was a clandestine nest under the bougainvillea. Seven ducks and two geese that I had ordered from Murray McMurray that spring were fattening in an open-air pen in the lot. Since they couldn't be trapped in a pen by an opossum or some other predator, I wagered that they would be safe, and a good source of protein. The young rabbits on the deck had gotten plump.

In my larder, I had jars of jam, stewed peaches, honey from last year's harvest, and pickles galore. My food-security future was bright. But as I assessed the food growing and thriving on the farm, a dark word crossed my mind, and I couldn't shake it. I walked upstairs and tried to forget. It felt like a gun to my head.

Carbohydrates.

I would have to ration the potatoes. Another potential problem crossed my mind, then another. Crop failure. Pests that kill plants and animals. Someone could steal all my food, an expansion on the great watermelon theft from the previous year. I peered out at the garden from my window. It was dark, and a wind had picked up. I could make out the cornstalks waving and the plum and apple trees rocking in the breeze. I felt a little queasy. As June evaporated at the stroke of midnight, suddenly my bold experiment, my attempt to prove myself as a farmer, seemed like the mission of an imbecile.

The next morning, as I picked a few apples to eat for breakfast, my first caffeine-withdrawal headache flashed across my temple. I had to go lie down.

Lying on my bed, with the morning sun filtering in and a breeze swirling the curtains into the sickroom, I wondered: How can I get out of this? It felt as if a monster had grabbed me and was going to hold me here for thirty— oh, no, why did I pick July?—thirty-one days. Why hadn't I weaned myself off coffee? Then another dreadful question: What's for lunch?

That afternoon Bill and I went to a friend's barbecue. Though I had eaten a jar of stewed peaches, a green salad, and at least ten ounces of honey, the smell of the grilling meat nearly knocked me down. Two yoga teachers I vaguely knew beckoned me over.

"I have the worst headache," I explained before they had the chance to read my aura.

"Give me your hand," Baxter said.

He pinched the area between my thumb and index finger. My headache went away. It was replaced by a growling stomach.

"I'm off coffee," I said with a sigh.

"You didn't do it gradually?" Raven asked.

I shook my head. Yoga people have been telling me for years that I should give up coffee, that it's full of toxins and other bad things. But when they suggest that I should stop drinking coffee, I want to tell them maybe they should saw off their legs.

Baxter gave me back my hand. The headache returned.

I looked around the party. There they were, my friends, standing next to the grill, dishing up salads, drinking beer. I had the sinking realization that social activities all revolve around sharing food. The act of setting up my 100-yard diet had turned me into an alien visiting from planet Weird in the solar system Healthy.

But then again, everyone at the party was on some kind of Bay Area diet kick anyway. The gluten-intolerant munched on ears of corn in the corner. The vegans had their own grill set up with toasting tofu. The raw-food vegans were sipping on freshly macheted green coconuts. The pescatarians were shoving ceviche into their faces. Defining ourselves by what we eat— that's what we do for fun around here.

I was sure that I could find a freshairian or a locavore to share my pain with but instead decided to leave early. I found Bill, an unapologetic omnivore, moving from grill to grill, stuffing sausages and ribs and veggie burgers into his mouth. I ripped a piece of watermelon out of his hand and insisted that, really, we couldn't stay another moment.

Later that day, I ordered three tea plants—*Camellia sinensis*—over the phone.

"I want the gallon size," I gritted out as the perky woman took my order. I needed a quick harvest.

"We'll include recipes for how to make tea," she assured me. The rest of the day passed in a painful haze.

On day two, I made several unfortunate discoveries.

With dreams of latkes dancing in my slow-moving, uncaffeinated brain, I made my way out to the garden with a shovel and a bucket. I have a half dozen potato zones in the veggie garden. One sprawled out of a neglected compost pile. I imagined the fat little crusters down below mixing with the dried-up leaves and stalks that had been breaking down over the years. A carbohydrate dream.

In February I had nestled the potatoes, organic blues, bought at the grocery store, at the bottom of the compost bin. Over their round shoulders I dumped fava bean leftovers, hay cleaned out from the chicken area, spent pea vines. As the green potato stalks emerged I bundled them with more straw and green matter. In *Matthew Biggs's Complete Book of Vegetables,* the British garden writer advised, "New potatoes are harvested when the flowers are blooming; larger ones once the foliage dies back." (He also mentioned that Marie Antoinette wore potato blossoms in her hair.)

I knew it was early, my potatoes hadn't yet blossomed, but Mr. Biggs had no idea how carb starved I was. A plate of mashed potatoes. If I could eat that, I would be happy for the rest of the day.

But now that I was digging, the plant, I had to admit, didn't look very healthy. I peered closer. Oh, no. Potato bugs, hundreds of them, were gnawing on the leaves and stalk. I plunged the shovel into the dirt and brought up a generous scoop. Grappling through the dirt, I found exactly two purple potatoes. Small ones. The size of marbles. The mother potato was deflated from this effort, and a few pale shoots slumped off of her girth.

I surveyed the rest of the potato plants tucked here and there around the garden. Instead of a seeing bountiful plants whose secret underground parts would get me through this experiment, I saw only unproductive freeloaders. What I had hoped was an iceberg of carbohydrates, with plenty down below, was reduced to an ice cube bobbing in a swimming pool. It would be a very small crop indeed. I carefully placed the marbles in my bucket and went upstairs to prepare my feast.

While the spuds fried in a dry cast-iron pan, I paced the living room, wondering what the hell I was supposed to eat for the rest of the month. During the Irish potato boom, people had plenty of food because potatoes grow easily—and, more than that, they make you feel full. Without carbs, satiety would be a distant memory.

Then I noticed our mantel. For the past two years, some corncobs I grew my first year of squat farming had lingered up there, along with a set of deer antlers and a white orchid plant. Indian corn, grown and saved for decoration. Once mere objects—now, as I gazed up at the multicolored cobs, I saw food. Carb food.

And so I did something I'd never done before. I ate an item of home decor. From a yellow-and-blue-checked ear of corn, I carefully pried out the individual kernels from their cobby home and piled them onto the table. As I loosened each kernel I felt like a prairie woman or an Indian squaw. I whispered thanks to my past self, the carbohydrate provider, who had thought to save those ears. One cob yielded a handful of corn. I deeply wanted cornmeal pancakes. But I didn't have a metate, the traditional stone grinder that Native Americans used, and I wasn't about to destroy my electric coffee grinder.

But I did have a Spong hand-cranked coffee grinder I'd bought a long time ago, out of nostalgia. It's made of metal painted black and red, with a little removable pan that catches the grounds. My mom's artsy friend Barb always hand-ground her coffee. Barb wore bohemian outfits (men's clothes, flowing dresses with skeleton patterns), had red hair down to her butt, and once had a pet crow. I remember visiting her kitchen in Idaho as a child. Barb and my mom flitted around the kitchen, laughing and glad to see each other again. My sister and I, standing on a chair, took turns grinding the dark beans for their morning coffee.

When, a few years ago, I spotted a similar grinder at a basement sale of an Italian imported-foods shop, I couldn't resist. And of course, I hadn't used it since. Who wants to labor, precaffeinated, over a hand-grinder for ten minutes in the morning? Yet now this grinder would be my salvation.

I carefully placed the kernels into the Spong. It was as if I were a kid again, standing on a chair and grinding. Only this time, I had my full weight propped up against the table so it wouldn't shake as I watched the kernels mill around in the hopper. But instead of the fetching aroma of fresh-ground coffee, I had the powdery residue of almost pure starch.

I've made cornmeal pancakes before, a family recipe adapted from *Joy of Cooking:* Add boiling water to cornmeal and let it rest. Add baking powder, salt, milk, an egg, and some melted butter, then mix. Of those ingredients, I had only the hot water and the egg. After letting an eighth of a cup of boiling water soak the yellow grain, I cracked an egg in, whipped it about, and poured the mixture into three blobs on a cast-iron pan.

I couldn't believe that the cakes actually puffed up like real pancakes. I ate them with a drizzle of honey and some stewed peaches on the side, with the blackened dwarf potatoes. They were the best pancakes I've ever eaten. I licked the plate. I counted the remaining corn cobs. Twelve.

❀

CHAPTER NINETEEN

July Fourth has always annoyed me. The endless gazing upward at a few flecks of light, the snarled traffic, people blowing off their fingers. But this year was different. This Fourth of July, I would be reaping the benefits of some work I had done months before.

In a wave of 1970s California nostalgia, my friend Jennifer and I drove up to Mendocino County last fall to pick grapes and make wine. Jennifer was a DIY lesbian who, when Bill and I arrived pale and eager from Seattle, taught us how to power our cars with biodiesel made out of fryer grease. Jennifer and I became friends and now worked together at the biodiesel filling station in Berkeley.

Jennifer had negotiated to exchange some biodiesel she had made for the grapes. When we drove into the valley, the vineyards were a riot of grapevines whose leaves were just starting to get their fall color. Purple fruit, the color of a bruise, hung amid green-gold leaves. The owner of the vineyard sat in his tractor. He was a tall, bearded hippie who grew biodynamic grapes. Jennifer handed him the jug of biodiesel with a look of triumph—she loved bartering. "The brix is at twenty-six," he said, referring to the sugar levels, and smiled. Then, in a rush because it was harvest season, he grabbed the jug of fuel and drove off, leaving us to harvest acres of grapes.

Days before, professional pickers had moved along the neat green rows and selected the best clusters, so we were picking the sloppy seconds. The overripe, the wrinkled, the tattered grapes left on the vine would become our wine.

It was hot when we picked, but the work wasn't hard. We snapped off clusters and dropped them into plastic lug boxes. The grapes were sweet and seedy. It only took an hour to pick hundreds of pounds.

The much harder work would be the crush, but luckily Jennifer and I had invited friends to help. Willow, always interested in gleaning and fermentation projects, had come over. First we pulled off the stems by hand, a circle of us gossiping and telling stories. Though there are crushing machines, we decided to do the crush in the traditional way. We poured the destemmed grapes into a large tub. Jennifer and I washed our feet and climbed in. There was a sickening moment when toes met grapes. Suddenly, it felt like we were standing in a pool of water. But we tromped and stomped. Our legs got sticky. It was kind of like an exercise machine. The party lasted well into the night. The yield: four five-gallon glass jugs full of grape juice.

Through the following winter and spring the juice bubbled and fermented in the jugs. Now, on the nation's day of independence, Jennifer and I would make the wine official by placing it in bottles and corking it.

I arrived at her place with only a slight caffeine headache, and we began bottling, using a tube, gravity, and some used but clean bottles. I've never been much of a drinker, but as I filled up bottle after bottle, I was glad that I had planned for the future. Putting up food is, at its heart, an optimistic thing. It's a bold way to say: I will be sticking around. Our wine had been fermenting for eight months. That's long-term planning for eating. Well, drinking.

And, in a way, bottling wine was the perfect way to celebrate America's independence. *The Alcoholic Republic,* by W. J. Rorabaugh, explains that a state of hunger and drunkenness was a way of life for early Americans, most of whom drank four ounces of distilled spirits every day. "The taste for strong drink was no doubt enhanced by the monotony of the American diet, which was dominated by corn," Rorabaugh writes. In the wild West, families subsisted on corn pone, salt pork, molasses, and whiskey. I, on the other hand, would be living on cornmeal, rabbits, greens, and wine.

While I happily contemplated spending the rest of July in a boozy torpor, Jennifer's roommates—amazing cooks—worked in the kitchen, roasting a chicken with new potatoes, pan-searing steaks, tossing salads. When everyone else took a break from bottling to eat real food, I wandered out to Jennifer's garden.

I tried to channel Euell Gibbons, the famous forager from the 1960s, whose books had been on the shelves of most ecologically minded folk of that era. His *Stalking the Wild Asparagus* is a beautifully written guide to harvesting cattails and milkweed pods. Nature provided; you just had to know where to look.

I knew the book because my dad is a big fan of foraging, and he had given me a faded green paperback version the last time I saw him in Idaho, about seven years ago. I had just reached the age of my parents when they started farming and I felt drawn back to the ranch. Bill, always game for a road trip, packed a spare tire and jugs of water for the ten-hour drive to Orofino, Idaho.

After a swim in the Clearwater River, which smelled just as I remembered—like swampy willow water but fast-moving and clear—we drove up to the ranch. I wanted to see the house my parents had built with their own hands: a rough-hewn cabin covered with cedar shingles and a tar-paper roof. I made Bill stop so I could pick some thimbleberries, berries in the *Rubus* family that my sister and I used to pick as children. They were velvety and tart.

The circular alfalfa field had gone back to thistles and small trees. The house had disappeared. Burned down. In the clearing where it had stood, the apple trees had gone feral.

After the disappointing visit to the ranch, Bill and I met up with my dad in town. I rarely saw him—only a handful of times the whole time I was growing up—but I could see that he, too had, gone feral. He smelled of woodsmoke and was wearing a wool shirt I had sent him years ago for his birthday. Bald, with a mustache, he walked a little bowlegged, but overall he was fit as a fiddle.

Over hamburgers in a diner in Orofino, he shrugged his shoulders about the house burning down. It and the property had been sold years ago, and he had given up the idea of being a rancher. He had been living in a small cabin without electricity or running water. He hunted for food, went fly-fishing.

When my sister and I were teenagers, our dad would send Christmas gifts of pine cones and photos of birch trees. These were worthless things to us—we craved the name-brand jeans we could never afford. But now, when

I think back on that, I realize that those were heartfelt gifts. He was trying to express who he was and what he cared about.

He handed me the book with one caveat: "Euell sold out," he said, and shook his head. "Goddamned Grape-Nuts." As Gibbons had gotten more and more famous, he had been hired to be a spokesperson for the cereal company. This broke my poor father's heart.

Now, as I stood in Jennifer's garden, I thought my father would be proud of me, foraging for my supper, living off the land as he does. I grazed on some red Russian kale, pulled a couple of green apples off the tree, and discovered a few Cape gooseberries—orange fruits that grow in tomatillo-like husks. They were sweet as honey.

I checked on Jennifer's bees, feeling a bit voyeuristic. Her healthy herd was finishing up the day's work; many of them loitered outside the entrance of the hive and dripped down the side of the box in a cluster. I was filled with longing for my own lost bees. I had tried to order another package, but the beekeeping supply stores were sold out. One man told me they sold out by January. Colony collapse disorder had hit beekeepers hard that year, so there were no surplus supplies for backyard beekeepers like me. Without bees, I had no honey.

I tried to ignore the good smells coming from the kitchen and went back inside to drink a few glasses of wine. A man at the party tried the substance we had made and declared it "a witty little wine."

By midnight, we had forced the last cork into the seventy-fifth bottle. Bill picked me up, and we stuffed my share—twenty-five bottles of not very good Sangiovese wine—into our station wagon. That it wasn't good didn't matter. The possibilities were endless. I could use it for cooking. I could make balsamic vinegar. Sangria. Mulled wine. As we drove home along MLK the festivities of the Fourth on our street unfolded: children holding Roman candles, a car that shot out a twenty-foot flame, police and fire engines roaring up and down the streets. I imagined that this is what our street would look like if there was a riot. It was wonderful. I was drunk.

CHAPTER TWENTY

B y day five, my headaches—and body aches—from caffeine withdrawal had subsided. I actually felt terrific. Light, energetic, with a thrumming, exuberant feeling from eating so many greens and salads and farm-fresh eggs. I rode my bike around, trying to remember the taste of the food I used to eat. I had pizza and Chinese food amnesia.

In the mornings, I would wake up and go to my feeding area—the garden. The new ducks and geese greeted me with great quacking shouts. They gorged down a few scoops of chicken food and nibbled at the bok choy from China-town I upended into their area. The geese ate first, always, and made a big show, craning their necks up and down—looking at me, then back to feeding.

"What are they?" someone called out from behind the fence. "Swans?"

"No, they're ducks and geese," I said. I peered between the slats of the fence to see a large woman with two children. A few of the ducks, following the sound of her voice to the end of the fence, stood and begged for food.

"Well, bless you and have a wonderful day," the big lady said. The chil-dren trailed after her.

I plucked an apple and a few plums, and made plans for lunch. The pumpkins were still small but numerous. I yanked the smallest from the vine; I'd shred it and make pumpkin hash browns. A red-chested hum-mingbird came down, letting out short bursts of air, then flew back into the ether. It was mating season for the little hummer; maybe he mistook me for a potential competitor.

I squeezed the green stalk of a corn plant. Still just ear and silk, no sub-stance. The Brandywine tomatoes were causing me much heartache. They were enormous but stubbornly green, and even after a hot day, they never threatened to blush.

I crouched near the zucchini plants and examined them. Beneath the giant turgid leaves, the fruits were still too small to eat. There was an abundance of the yellow-orange flowers, though. I had heard you could eat them, so I gathered a colanderful, with the intention of frying them.

I gave the flowers a quick rinse in the sink and shredded the pumpkin with the grater. When I dabbed the flowers dry with a tea towel, I heard a curious noise. A muffled buzz. I checked my cell phone: no. It was coming from the flower. I peeled back the crepelike lips of the zucchini blossom, and out veered one very upset fuzzy black bee. It adjusted to the new light and hastened toward the open back door. My heart beating very fast, I picked up another flower and pried open the blossom. Another furry prisoner buzzed out. Four captives were released before I could eat lunch.

I dredged each of the flowers in egg, dipped them in cornmeal, then fried them. I sprinkled the former bee prison with lemon juice and stuck it in my mouth.

After eating "lunch," I made plans to go to Willow's garden to harvest lettuce for some former Black Panthers. A few months before my experiment in self-sufficiency began, I had encountered the organization, which I thought was long dead. "Join the Commemoration Committee for the Black Panther Party!" a kid with dreadlocks shouted outside North Gate Hall on UC Berkeley's campus, where I was taking some classes. He stood behind a table with another man, behind a stack of newspapers with the words THE COMMEMORATOR and an image of a black panther busting out.

What did I know about the Black Panthers? Black power, guns, men in leather jackets, Huey Newton sitting in that big wicker chair. My mom and dad were active in the civil rights movement and had lived in Berkeley and Oakland when the Panthers first started in West Oakland in 1966. I was curious and wandered over.

The two pamphleteers were in the middle of explaining to a young black student that the Black Panthers were necessary for social justice in America.

That education was the most important thing for all the kids who lived in the inner city. I nodded my head in agreement.

"Can I help?" I asked the kid and the middle-aged man standing next to him. Remembering the scene from the Malcolm X movie where the blond lady's help is rejected, I figured they would say no thanks, whitey.

"Yes, we're integrated now," the man said, and handed me the Black Panther Party Ten-Point Program.

"Well, I don't have time or money, how about vegetables?" I asked.

The man, whose name was Melvin, beamed. "We could really use some salad for our literacy program."

Melvin took down my name and promised to call. I read the Black Panther Party Ten-Point Program in my car. The list covered employment demands, an end to police brutality, and education and health-care concerns for "our black and oppressed communities." The final program point read: "We want land, bread, housing, education, clothing, justice, peace and people's community control of modern technology." It was followed by the first two paragraphs of the Declaration of Independence, the document that embodied the ideas of the American Revolution.

When I called Willow and asked her about supplying salad for a literacy program—I wasn't sure I could grow the volume of lettuce necessary alone—she told me that her nonprofit garden project had been inspired by the "survival programs" of the Black Panthers, in which they distributed food and eyeglasses to the needy. "Hell," she said, "we'll plant a Lil' Bobby Hutton memorial plot of lettuce!"

Every week since that meeting, I had been harvesting lettuce from both my garden and City Slicker Farm to share with the Black Panthers' literacy program. For the month of July, though, I couldn't share my bounty. So, after my lunch of fried flowers, I swung open the gate to one of Willow's community farms and yanked five sturdy-looking heads of lettuce out of the ground. I snipped off the roots, leaving them there on the ground to rot back into the earth, and tucked the leaves into my bag. I chose the red frilly Lolla Rosa, the bright green Deer Tongue, and Speckles, a green lettuce with red spots.

After washing and bagging the greens at my house, I got on my bike and rode through GhostTown to deliver the lettuce to the office of *The Com-memorator*, the newspaper of the Commemoration Committee for the Black Panther Party. It was the errand of an optimist. I knew that providing a salad once a week to kids at their drop-in literacy program wouldn't change anything. But I did it anyway because—if I'm honest with myself—it made me feel better. It gave me hope.

I pedaled up Martin Luther King Jr. Way on my ten-speed, a bag of salad greens gently rocking on the handlebars. I spotted Johnny, the Watermelon Man, who sells watermelon in the summer and greens in the winter at his produce stand. I have never noticed much buying going on; mostly he and some other old-timers just hang out under the awning of the little shop. As I pressed on I noticed a man sleeping in an overturned refrigerator box, arms flung out like a baby. I counted five men and one woman with shopping carts filled with aluminum cans headed to the recycling center.

Acts from people's lives are played out on the streets and sidewalks like Shakespearean drama. On this July day, whole families sat on the sidewalk, chairs placed just so, to take in or be part of the day's events. Just the night before, I had happened on a woman yelling at the father of her son for money he owed her. While she ranted (from the seat of a Hummer), his friends recorded the performance on their cell phones. "You're acting like some kinda Michael Jackson," she hurled, and the Hummer screeched away. The man and his friends cried out at this dis, slapped street signs, and groaned.

A person on a bike gets to be part of this sidewalk theater. I got a sweet "Hello, good morning" from a man walking with a cane across the crosswalk.

After thirty flat blocks, the landscape changed. I crossed the border into Berkeley. There's a NUCLEAR-FREE ZONE sign and the giant, gleaming words THERE and HERE, a lumbering piece of public sculpture that has always rubbed me the wrong way.

"There is no there there," Gertrude Stein once famously said. Though she was referring to her Oakland childhood home, which was destroyed in a fire, seventy years later Berkeley, in the form of public art, continued the misunderstanding that she was dissing all of Oakland.

The gold-embossed sign above the door read THE COMMEMORATOR. The quarterly paper has a circulation of ten thousand and specializes in stories for the black community. It was started in 1990, a year after founding Black Panther Huey Newton was killed. The remaining members of the party felt the Panthers' socialist legacy might be lost.

I rang the doorbell, and Melvin Johnson, tall and handsome yet bogged down by the formidable task of running a newspaper, let me in. The office smelled of incense, and on the wall there were hand-painted drawings of various Black Panther all-stars. There were Alprentice "Bunchy" Carter and Mumia Abu-Jamal and an oil pastel of a Black Panther youth group, all painted directly onto the wall.

"Hey, salad lady," called the other Melvin, Melvin Dickson, a stocky, muscular man with kind eyes. Dickson was an original Black Panther, in charge of all things culinary for the Bay Area Panthers from 1972 to 1982. After I put the lettuce in the fridge, we often sat in the office and chatted about events and history. I found myself frequently asking him for advice.

"I see kids eating all this junk food in our neighborhood," I said the first day I dropped off the lettuce. "That's why I'm bringing this salad."

"Kids are hyper on that junk food," Dickson said. "They can't learn in that state of mind. One thing we imparted was a nutritious diet. That's why we fed them three meals." The Black Panthers weren't just about guns and self-defense; they started a free breakfast program for hungry children. Later, in some of their schools, they served breakfast, lunch, and dinner to the students, so their parents could go to work. I thought about how different my neighborhood would be if those self-sufficiency programs had survived.

When I wished aloud that more programs like the ones the Black Panthers started existed today, Melvin sighed. "There just aren't any programs anymore. You've got to challenge them, educate them, get them to try new things."

I knew Melvin was right, but now that I was surviving on lettuce and pumpkins for several days, I do believe I would have killed someone for a bag of red-hot Cheetos.

CHAPTER TWENTY-ONE

❀

On day ten of the experiment, I stood on the boggy roof of an abandoned carport eating plums. The house was abandoned, too. The tree, planted at the back of the house by some kindly farmer of yesteryear, groaned with fruit.

In order to be truly self-sufficient for the month of July, I found that I had to become a hunter-gatherer of sorts. There was no shame in this—I couldn't grow everything, after all. Even Wendell Berry, farmer extraordinaire, agreed. In the essay "The Whole Horse," he wrote, "A subsistence economy necessarily is highly diversified, and it characteristically has involved hunting and gathering as well as farming and gardening." It was true that eating the same things out of the garden—lettuce, beets, squash blossoms—day after day had gotten a little monotonous. I needed to supplement with some foraged food. According to Roman law, it is perfectly legal to harvest fruit that hangs over into a public area.

I spotted the plums while I was riding my bike. I had never noticed them before, but the 100-yard diet had so heightened my senses, I started to see food everywhere. Every shrub, tree, and weed I encountered quivered with potential usefulness. In every abandoned lot, I saw a potential garden. I could also smell a hot dog a mile away.

These plums were a variety called elephant hearts. They had green skin and bright red flesh, in the shape of a heart. They didn't taste particularly good. In fact, if I hadn't been doing this experiment in self-sufficiency, I never would have gone out of my way to find the tree, shimmy up a wooden fence, make the catlike leap to the garage, and creep across the rotting beams for a few plums. And now that I had gone through that, I found them to be vaguely dry, maybe too sour. But I was hungry, so I scarfed them down on the rooftop.

As I munched, I silently thanked the long-gone home owner who had planted this tree. Whoever had done so probably had to make a tough decision: a beautiful ornamental or a fruit-bearing tree.

"Garden style is a continuing expression of the changing idea of the universe," environmentalist Paul Shepard observes in *Thinking Animals,* pointing out that Italy's Renaissance gardens were orderly and complex, like aristocratic society. If this is true, and I think it is, what does our city landscaping say about us? The barren ornamental pears, the trimmed hedges, the ubiquitous lawn—the pedigreed landscape. I find this environment to be wasteful. "The observer of city gardens cannot fail to notice that not one of the plants that are grown in most urban residential areas, or that appear on planting plans, have the slightest nutritional value," landscape architect Michael Hough writes in *City Form and Natural Process.* "However, opportunities for using edible plants are just as great as [for] using those that are purely ornamental. Tree planting along city streets could include fruit-bearing species."

Here, someone had ignored convention and planted this fruiting plum tree. Maybe he had been hungry. Maybe the tree reminded him of home. Maybe he had imagined plum dumplings or plum jam. Whatever his motives, he watered the tree, didn't cut it down, let it flourish and fruit for all these years. Based on its size, it must have been forty years old. Whoever planted it could never have predicted my existence—a crazy, starved, foraging locavore. The past was feeding me today, and I was grateful.

After I finished eating, I loaded two plastic bags with fruit, let them fall to the soft earth, and climbed down after them. I balanced the bags on my bike's handlebars and headed home. I had a hunch. It involved canning.

On my way, I paused a few blocks from the 2-8 to watch a dice game. Two boys were playing—one fat, one thin. They yelled and rolled. The fat one threw down a dollar.

"Excuse me," I said. "How does this work?"

Without pause, as if he had been waiting for someone to finally inquire, the thin kid explained that the first roll determines the bet. If it's a seven, for instance, then the person who bets is betting that another seven will be rolled.

I watched for a while, and the fat kid lost all four of his ones.

"Can I have 'em back?" he said to the thin kid.

"OK." The skinny kid passed him the floppy bills. I wasn't the only one just playing. This kid was pretending to bet; I was playing at self-sufficiency.

I continued cycling, keeping the BART tracks and highways 980 and 24 on my left.

I passed the lumbering *Magnolia grandiflora*s growing along some of MLK. The trees have leathery leaves and giant white blossoms, and if it's not rush hour, you can smell their tangy-sweet lemon scent.

At home, a woman who had recently moved to our street, Makeda, was in my garden. She, like all of us, has a hustle. She makes pulled-pork sandwiches, stacks them into a wheelie cart, and then wanders around Oakland's small rock-club-and-bar district after dark selling them.

"Hey, Novella, can I pick some beets?" she asked, her red dreads glimmering in the sun. She had asked to pick beets before, and I was always more than happy to share. But now that I was on this garden-eating stunt, it felt like she was asking for my firstborn.

A stray cat was in the garden with us. Gray, lanky, and half wild, he usually ran away at the first glimpse of a person, but he was so intent on stalking a mouse in the compost pile that he hadn't noticed us.

"Sure, sure," I said to Makeda, and showed her which ones to pull from the dark earth. I had to stay human, I reminded myself as I parked my bike.

Upstairs I dunked the plums in a bucket filled with water and mercilessly scrubbed them down. I loaded my oven with widemouthed jars, and boiled water in a giant blue enamel canning pot. After the jars were sterilized—really hot—I crammed as many whole plums into the jars as could fit. I boiled the jars of plums in the water bath—this process is called raw-pack canning—and once some of the plums had softened and cooked down, I crammed in a few more until they were an inch from the top of the jar.

Then I screwed on the lids and let the jars rumble under two inches of boiling water for about an hour. When I pulled the jars from the water, the plums had turned an amazing fuchsia color. I placed the hot jars of plums into our pantry to cool down overnight and set the seal.

That night, Bill and I went out to an East Bay institution we had heard about but could not believe until we saw it with our own eyes: in the parking lot of a bakery, four metal bins overflowing with loaves of bread, any time of day, any day of the week.

Day-old loaves. Bread that was too dark, too pale, or otherwise damaged in some way in the bakery was dumped. Pastries not sold, also dumped. And so these Dumpsters attracted and nourished the entire scavenger community of the East Bay: the hippies, the punks, the scroungers. Occasionally regular citizens appeared, gawking at the plenty before sheepishly snagging a few loaves. These were the Dumpsters of Life.

The smell of the bakery almost knocked me over. Behind a glass wall, men wearing white surgical smocks sweated, forming dough, mixing flour, pulling pans out of the oven. I stared at them in awe. Carbohydrates in action. So delicious, so not allowed according to rule number three: No food from Dumpsters (except to feed the animals).

The bread Dumpsters were in the back of the parking lot. They were big, lumbering, each the size of a minivan. ANIMAL FEED ONLY, one sign read; NO TRESPASSING, another read. Bill and I, now veteran Chinatown scroungers, were not shy. We flipped the black plastic lids back; they clanged against the green metal of the Dumpster. The smell of bakery items was intoxicating. I took a deep huff. Bill immediately snagged a cinnamon roll that rolled freely on top of a pile of challah, baguettes, and liberated slices of bread. He gnawed on it thoughtfully, then concluded, "Mmm, good." I was both repulsed and desperately jealous.

I was there for the rabbits.

Riana informed me that Mamie, her French grandmother-in-law, always fed her rabbits stale baguettes. I couldn't find a source in English that recommended such a practice, but I was eager to save money. The alfalfa pellets were starting to get expensive. I hadn't been spending any money on food for myself, so I was actually saving tons, yet I was keen to explore the city's bounty. And there, in the Dumpster of Life, within easy reach, were twenty-four baguettes. We hustled them home, along with some hamburger buns for the chickens and five cinnamon rolls for Bill.

The rabbits fell on the stale baguettes as if they had been waiting for them their whole lives. They ignored the bok choy and sharpened their teeth on the hard bread.

The next day for breakfast, while Bill heated up his unbearably delicious-smelling Dumpster rolls, I opened a jar of the stewed plums. Just as it should be, the lid was tight and hard to pry off, but it finally yielded with a satisfying pop. Inside, a thick juice the color of wine covered the plums. I took a swig. Sweet, thick nectar with a slight hint of cherry filled my mouth. I dug into the flesh of the plum on top with a spoon. It was dense and puddinglike, tart but not as sour as the raw fruits.

As Bill munched on his roll and I ate a whole jar of plums I wondered how the bread-fed bunnies would taste.

CHAPTER TWENTY-TWO

Finally, after one missed UPS delivery, the tea plants arrived. The caffeine-withdrawal headaches were now gone, so it was with only mild interest, not desperation, that I opened the long cardboard box. The three plants were wrapped in butcher paper; once released from their brown swaddling, they looked terrifically healthy, like shiny-leaved, ornamental camellias. The invoice reminded me that I had paid $20 for each plant, so I had to put them to good use.

Two of the *Camellia sinensis* plants had new bud growth, which is what's usually picked to make green tea. I planted them in the front yard, in a semi-shady area, and took a few young leaves upstairs. According to the instructions that came with the plants, green tea is the easiest to make. You simply pan-fry or steam the leaves, then dry them out. Within a few hours, I had some grassy-smelling green tea. It wasn't coffee, but at least it wasn't another mug of mint tea. A few minutes after drinking it, I felt a surge of energy and well-being. Probably the best $60 I ever spent.

Bill, a huge fan of green tea—jasmine-spiked was his favorite—came into the kitchen as I brewed my second mug.

"Can I taste it?" he asked, reaching across the table to snag my mug.

I made a grunting sound and grabbed the mug back, careful not to spill a drop.

"You get to eat whatever you want!" I said. I had been coveting his breakfasts of cereal and milk.

"Just a taste?" he begged.

I surrended. He took a huge, slurping gulp. "That's awful," he said.

"More for me," I said, and reclaimed my cup of caffeine.

Near the middle of my experiment, I noticed that one of my hens had gone broody. A broody chicken will sit on her egg—or, if you haven't been gathering them daily, the clutch of eggs—and will refuse to move. A broody chicken is an intense animal, devoted to, obsessed about, hatching some eggs. This lasts for about three weeks, the usual gestation time for chicks. Even if there's no rooster and no chance for a baby chick (save for an immaculate chicken conception), the hen still sits firmly on her nest and does nothing, not even laying more eggs.

I had been eating—and depending on—up to three eggs a day to keep up my protein levels. I went downstairs to consult the chicken. I brought a cabbage leaf. She had set up a nest near the side of the house, under a bush.

"Hello?" I started.

She made a horrible keening noise.

I went to pet her feathers. She pecked me, hard.

I offered her the leaf. She stared intently into the middle distance. She seemed annoyed at my audacity, my wilted bribe.

I prayed that the other chickens didn't go broody. Experts I consulted on the Internet said there's nothing really to do about a broody hen; you just have to wait it out.

As I calculated my protein intake a duck walked by. He was one of seven I had been raising, a white Pekin, like the ones our little neighbor Sophia had loved, one of which had been killed by the opossum. Sophia and her mom had moved away that past winter.

The duck gave me the hairy eyeball that ducks tend to give, cocking their heads up, beady eyes wary but charming. I bought a pair of pruning loppers the next day.

The previous batch of ducks, the two who had survived the opossum attack, had ended up on my table around the time Sophia and Neruda moved away. I didn't have the heart to kill them personally, but they were messy and were eating tons of feed, so I hired two assassins. "Assassins" might be a strong word. More like two hungry hippies Bill knew. I watched, like a coward, from the kitchen as they carried the ducks from the back porch and into the lot, where they chopped off their heads with an ax.

It took an hour to pluck the ducks, and then we barbecued them. The dinner was the hippies' payment. Alas, the meat was hard and rubbery because we hadn't let it rest. The skin and fat were delicious, though.

One of the hippies chewed thoughtfully. He was tall, had long hair, and frequently went barefoot. Around his neck he wore a rope with a bottle of gin tied to it. On the other end of the rope was a bottle of tonic water.

Where do you find these people? In Oakland. "You know, modern man doesn't get to use his teeth much anymore," he pointed out. "This is exercise for my teeth!"

I remembered that Carla Emery's *Encyclopedia of Country Living* had suggested resting meat before eating it to keep it from getting rubbery and tough, but I didn't understand why. For a clear, scientific explanation, I turned to Harold McGee's encyclopedic *On Food and Cooking:* "For a brief period after the animal's death its muscles are relaxed and if immediately cut and cooked will make especially tender meat." Hmm. The hippies and I probably took too long to pluck the ducks. Rigor mortis had set in, the protein filaments in the muscles bound together, creating a tough texture.

If we had let the ducks rest for twenty-four hours, according to McGee, enzymes called cathepsins would have broken down the bound filaments, making the meat tender. The enzymes also break proteins into tasty amino acids and fats into aromatic fatty acids. "All of these breakdown products contribute to the intensely meaty, nutty flavor of aged meat." Oh, hell, I thought, what a waste.

This time around, in a state of semistarvation, I went out to the lot and grabbed the first duck I could catch. I didn't want to kill him in front of the other ducks and geese, but the backyard was occupied by chickens who might take offense, and on the deck were rabbits who would certainly become upset about an execution. So I took the white duck into our bathroom and plopped him in the tub with some water in it. He quacked and swam around for a few minutes while I collected my arsenal: a bucket and the recently purchased tree pruners. A friend of mine who keeps ducks

kills them using this method, which he calls harvesting. It wasn't like killing Harold. I merely opened up the loppers, placed them around the duck's
neck, and squeezed the loppers shut. The duck went from being a happy
camper to being a headless camper. I plucked and eviscerated him outside
on a table. The killing thing was starting to feel a bit routine.

After the duck rested for a day in the fridge, I baked him with the oven
on low, letting his fatty skin baste the unctuous meat. I decided to share a
little of the duck with Bill. The meat was tender and delicious. Not used to
so much food at once, I paused to digest and watch Bill eat. As he gnawed
on a duck breast, his lips and chin growing greasy from all the fat, I was
reminded that we really weren't so far away from monkeys. Chimps eat
meat—eliminate the concept of the banana-loving fruitarian—and meat,
according to Susan Allport's compelling book *The Primal Feast: Food, Sex,
Foraging, and Love,* "is the food that is most often fought over, stolen, begged
for, and shared."

Only slightly paranoid that Bill would find them, I put the duck leftovers
in the back of the fridge. The fatty skin became my version of bacon for a
few days.

CHAPTER TWENTY-THREE

❋

I pulled my honey extractor out of the car and lugged it across Jennifer's yard. Past the fava beans, past the stout kale soldiers, toward the apple tree. Jennifer, just as I had hoped, felt sorry for me in my beeless state and had invited me over to extract some honey.

When I reached the apple tree, I suddenly noticed a strange hum filling the air. Jennifer stood behind her beehive, and when she saw me, she said, "It's happening again."

Swarming is a perfectly natural event in the hive. When the worker bees start to feel that their quarters are cramped, too filled with honey, or simply unsuitable for a variety of reasons that human beings don't understand, they make arrangements to move out. First they construct one or several peanut-shaped queen cells on the outer areas of the frames. Then they place a worker-bee egg in the cell. The workers feed the growing bee royal jelly, which contains hormones that create a queen. When the queen hatches, a group of the worker bees departs with her for new digs.

Jennifer and I now had front-row seats for this dramatic departure. Each bee that had volunteered to join the new queen flew out of the hive and began flying in a circle around the yard. At first it was just a few bees, then more, then more, until the air was thick with them. Jennifer and I stood very still. As the minutes went by the circle got wider, so that every inch of Jennifer's backyard seemed to contain a bee. Somehow they avoided us, two dummies standing still next to a stainless-steel extractor. The noise was unbearable. I noticed I was holding my breath.

I wasn't scared, but it was the same feeling I got when I opened up a bee box for an inspection. You must be fearless. You retreat to a strange, calm, empty place in your mind. Now that I thought about it, I had retreated to that place when I was almost-mugged by the gun-toting kid, and when I was finally able to walk around our neighborhood. Was it facing fears? Giving myself up to the fate of the universe? The bees had definitely taught me how to let go.

At some magical signal, the bees fell in together, flew in a cartoonish pack toward a lemon tree, and landed on a low-hanging branch. It was my lucky day. If I could catch the swarm, I could become a beekeeper again.

The only problem was that the lemon tree they had landed on grew in the neighbor's backyard. And the neighbor's house was for sale. And there was an open house going on right at that very moment. I peeked over the fence and saw the blond, lipsticked real estate agent looking at the throbbing swarm with an expression of utter horror.

Though they are intimidating, swarming bees aren't dangerous at all. They don't have anything to defend, and they are too enraptured by the scent of the queen to sting anyone. Queen pheromone is the stuff that beekeepers and exhibitionists use to create "bee beards."

Fearing that the agent might try to annihilate these sisters of goodness and light with some insect killer in order to sell a house, I vaulted over the fence and walked close to the swarm.

"See, no problems!" I tried to communicate with body language. I waved to her. Jennifer snickered behind the fence. Time was of the essence. Once they settled into the tree, scout bees from the swarm would fly off to explore possible new homes. I figured I had an hour.

I rushed to my home in a state of total panicked glee, driving the two miles like a madwoman. If the bees left the lemon bough and flitted off to some inaccessible place, they would go feral, return to nature—and I would remain beeless. I screeched to a halt outside my door, left the car running, waved at a surprised Mr. Nguyen, thundered upstairs, pushed past the rabbits on the deck, and grabbed the vacant bee box, which hadn't moved since the day of the drug bust.

Back at Jennifer's within ten minutes, I saw with relief that the swarm hadn't moved. Jennifer set up a ladder into her neighbor's backyard. I schlepped the bottom box over the fence and set it under the tree. Then I tried to entice the bees into this honey-scented place.

This involved a stick. I gently prodded a few of the bees down into the box, but they flew right back to the cluster. Glancing over my shoulder to make sure the real estate agent or potential home owners wouldn't disturb me, I started firmly shaking the tree. A clump of bees tumbled into the box, crawled around, then returned to the cluster. The tree was a small one—it looked like a dwarf Meyer lemon. The branch they had landed on was too big to cut with pruners. It would require a chainsaw.

While I contemplated who might have a chainsaw and considered the fact that bees hate the sound of engines, I gave the branch another shake. A big chunk of bees fell into the awaiting box. The queen must have been part of the chunk, because suddenly all the bees abandoned the branch and cascaded into their new home. Having caught them, I triumphantly slapped a lid over the box and stuffed a T-shirt into the hive entrance. Then, feeling like a pirate, I clattered up the ladder and climbed over the fence with my new treasure.

Like a bad parent, I left the bees in the car with the windows rolled up as I went back to extract honey with Jennifer. I reasoned that they would be fine—as I worked, they would be crawling around their new home, making plans for redecorating. The former occupants had left a good foundation for building a new colony: Most of the frames had pulled honeycomb—the wax cells within which the bees store honey and pollen and house their young— some pollen, and a little bit of honey.

Bees expend an enormous amount of time, energy, and nectar on making pulled comb. They have a feedback mechanism that determines when to make honey or wax. After a day of foraging for nectar, the field bee transfers her load, via her tongue, to a house bee. This bee then ferments the nectar into honey. However, if there isn't enough comb in which to place the honey, she just digests the nectar longer, creating wax. It's an elegant and simple feedback loop.

Back in Jennifer's yard, while the bees hung out in the car, we extracted Jennifer's honey. That her bees were swarming meant that her hive was too full of honey. We spun out two full supers of honey. Since I was helping and loaning her my extractor—which had become a communal device, shared by Willow, Joel, Jennifer, and me—I was awarded two quarts of honey—a huge bounty for a 100-yard dieter.

As I drove home I felt a slick of sweat on my back from the hard work of extracting, the hot day, and nerves. This was my first swarm capture, and I was giddy. I looked into my rearview mirror at the thrumming white box and spoke to the bees about my garden and the merits of their new home. I felt a little proud of my bees. They were, after all, taking a risk by leaving the comforts of their safe, honey-filled home. Why did these particular bees choose to leave while others stayed? There are some things we don't know about bees. My wager was that they simply liked the way this new queen smelled. They were like immigrants, following a new future.

Once we were home, I proudly placed the box on a scavenged old-fashioned schoolchild's desk in the middle of the squat garden. I was tempting fate and the owner of the lot, but I knew the bees would be happier down in the garden. Plus, they probably would have stung the new resident rabbits on the deck. I removed the T-shirt from the hive entrance and watched happily as the golden specks investigated their surroundings, landing on my carefully tended garden, then flying into the sky for a better view.

A few days later, I walked by Brother's Market on my way to get weeds for the chickens, ducks, and geese. Bill and I hadn't gone Dumpster diving lately. Mosed, the shopkeeper, was outside enjoying the sun. When he saw me, he pointed at the buckets. "What do you do with those?" he asked.

"They're for my chickens," I said.

"Where are these chickens?" he asked, sure that I was quite insane.

"At the farm down there," I said, and pointed at the lot, which you could see from Mosed's doorway.

"Where?" He peered.

"Come on," I said, "I'll show you."

As we walked the half block to the garden I explained that I had chickens, ducks, bees, and rabbits. His ears perked up when he heard me say bees.

"In Yemen, we have the best honey," Mosed declared. "The best beekeepers in the world are Yemeni."

I nodded. Most of the liquor-store owners in GhostTown and West Oakland were from Yemen. I was told they were able to avoid being robbed because it was well known that they were armed to the teeth. And, indeed, a few months later, when a drive-by shooting occurred at Brother's Market, Mosed and the other shopkeepers returned fire.

"You'll have to try our honey," I said.

"Where are you from?" he asked suddenly. He didn't mean which town or region in America, he meant my roots.

Usually white people like me don't have to talk about "where they come from." I wanted to answer his question, but I wasn't sure: I'm an American; I'm from nowhere and everywhere.

I thought of my mom—a liberal ex-hippie born in Newport Beach, California, maiden name Schultz, who now lives in the Pacific Northwest. My dad—a hermit from the wilds of Oregon who now lives in the woods in Idaho. All my grandparents were dead; I had barely known them. I was a slurry of German, Norwegian, and French stock. The question stumped me. I shrugged.

Mosed and I reached the garden, and the ducks and geese erupted into a burst of enthusiastic quacking. Their run ran along the fence next to MLK, and they usually spent their days calling out a similar greeting to all manner of passersby. I planned on mating the goose and the gander. I hoped to mate the ducks—white Pekins—with Willow's Muscovy ducks, the goal being Moulards, the ideal cross, popular in France for making foie gras.

Mosed seemed noncommittal, certainly not impressed by my ducks or the garden. But his eyes lit up when he saw the beehive. A cloud of bees hovered near it, some coming, some going. While he watched their labors I brought him two kinds of honey from our cupboard.

The fall honey, harvested from my bees before they died out, was dark brown and tasted a bit of eucalyptus. "We call it red honey," Mosed said, and sampled it. "And this is the white honey," he added, pointing to the thinner version—what I called fennel honey—which Jennifer and I had harvested a few days before. He liked the red honey very much and offered to buy some. But the 100-yard-diet experiment was fast depleting the honey supply, so I had to say no—I didn't have much left. I promised to sell him some from the next harvest.

As I finished the tour of the garden Mosed pointed to a fava bean plant. "We call this *yaell,*" he said. I yanked up the stalk, heavy with fat green pods, and handed it to him.

"Thank you," he said, and hugged the plant.

Mosed disappeared down the street. He expressed an interest in some of the chickens, for meat. It was important that he got the chickens live, so he could kill them as the Muslim tradition dictated. I nodded and promised to look into it.

I didn't know what I was or where I was from, but I did know that I would be happy to be Mosed's farmer.

❋

CHAPTER TWENTY-FOUR

On day seventeen, in the car, Bill made an announcement. "I don't want you to get upset, but ever since you started this experiment," he said, "well, your breath stinks. Really stinks at night." I had just picked him up from the BART station, and he had moaned that he was hungry, so I agreed to go to our favorite El Salvadoran restaurant just to keep him company.

"What does it smell like?" I asked as I focused on the road. We had just been talking about how it's good to admit mistakes, and I was trying to take this new, somewhat unpleasant information in stride.

"I don't know, I can't describe it," he replied.

"Sour? Like death?"

"Kinda dry-like."

I'd wondered why he hadn't been very cuddly lately.

"Maybe you should brush your teeth," he said.

"I do brush my teeth!" I said. Toothpaste was allowed in the 100-yard diet.

"Oh. Well, don't get mad."

I wondered what could be causing this halitosis. If Bill said my breath stank, it must. He's a low-maintenance guy who rarely brushes his teeth and washes his hair with bar soap.

Coffee—it had to be the coffee missing from my diet. "Maybe the acids from the coffee kill off the bad bacteria in my mouth," I said. I really missed coffee.

"I don't know." Bill had been against coffee ever since he had quit two years before and started drinking green tea instead.

In addition to the halitosis, the experiment was taking a toll on how we

spent time together. As with most couples, an intrinsic part of our relationship was eating—foraging, dining out, and cooking together. The sad fact
was that the 100-yard diet was tearing us apart.

We used to eat out about three times a week, both of us exhausted after
work and too tired to pull vegetables out of the garden to cook. I missed
this connection, so I agreed to watch Bill eat at Los Cocos. The place was
a hole-in-the-wall in a mostly Latino part of town. Little ladies patted
corn masa full of cheese and beans, then slapped it onto the grill. Within
moments, a delicious fat tortilla full of runny stuff—a pupusa—was served.

That night, the place was packed, and all the customers were grouchy
and hungry. The little ladies behind the counter looked stressed out. Not
being part of the exchange, I observed the place in a new way. It all seemed
like a ruse—the tables, the chairs, a made-up world, a piece of theater.

Bill didn't get his food for half an hour. I sat and watched the process
of the restaurant as if I were in the bleachers of a tennis match. Someone
ordered, sat down, ate chips, slumped over. When the food arrived, they
gulped it down. I remembered the feeling. You're so hungry you can't enjoy
the food—you're just fulfilling a bodily urge.

Since I was subsisting mainly on grated pumpkin, stewed plums, and
a steady dose of wine every night, I wanted to tap on the shoulder of the
man in the corner who was done eating but had left half of his pupusa on
the table and tell him how lucky he was to have enjoyed something so com-
plicated and, from what I could remember from past visits, so delicious. I
would point out that the corn that had made the masa had been pulled off
the stalks, then ground into precious cornmeal. Or the beans, the glimmering black beans, had been threshed, gathered together, then stewed for hours
in a pot, with a generous helping of lard, no doubt. "Cows have been milked
for your meal," I would shout, "so finish your food!" Yes, I had become
totally obnoxious. Luckily I managed to keep all these thoughts to myself.

Restaurants weren't my only problem. The Dumpsters were killing me.
The Chinatown green bins, which Bill and I visited twice a week, were start-
ing to smell extremely good to my food-deprived body. If I found an apple,
I would pause to stare at it for several long minutes, trying to figure out why

someone had thrown away a perfectly good piece of fruit. This apple—cousin to those I had been plucking from my own tree—had started off as a blossom; a bee had landed and fertilization had happened; then the fruit had ripened through the spring and summer, until it was picked, washed, sorted, and shipped to the store. There it sat on the shelves, caressed by many hands, until it was tossed in the Dumpster. A baffling trajectory. Then I would hold up someone's discarded takeout container and marvel at the full portion of kung pao chicken lying in a pool of orange goo. The goo looked so tasty, so forbidden. Why was so much food being wasted?

As I grew dizzy in the pupuseria thinking about the intricate, wasteful food spiral in which we all take part daily, I remembered M.F.K. Fisher's *How to Cook a Wolf,* about cooking during the food rationing of World War II. After Bill shoveled down his beans and rice, we went home, and I dragged out my copy of the book.

I read the introduction a loud to Bill: "Butter, no matter how unlimited, is a precious substance not lightly to be wasted. . . . And that is good, for there can be no more shameful carelessness than with the food we eat for life itself." I paused for dramatic effect. "When we exist without thought or thanksgiving we are not men, but beasts," I finished triumphantly, trying not to direct my bad breath at Bill. He rolled his eyes.

My bad breath, my righteousness, my unwillingness to share—after only seventeen days, the 100-yard diet was really putting a strain on my relationship.

The next day, from my living room window, I watched a man shoot up. He hunched over in the vestibule of the abandoned brick building. It's generally an area where people dump items they don't want, and it sometimes becomes an impromptu bathroom for the desperate. I hadn't seen it used as a shooting gallery before. The guy wore a hooded sweatshirt and sat on a bucket.

He stood, pulled his pants down, then sat back down. I was queasy. I suddenly felt ashamed about my foodie righteousness at the restaurant. There *is* something more shameful than carelessly eating or wasting food: wasting people. All the crackheads and the prostitutes, the junkies and the homeless, in my neighborhood—they are evidence of far bigger problems than mere nourishment.

To the chalkboard tally—

25 rabbits
4 ducks
2 geese
4 chickens
10,000 bees
68 flies
2 monkeys

—I added: 1 junkie.

When I was done writing, I peeked out the window again. The druggie was still out there, sitting with his pants down, asleep.

CHAPTER TWENTY-FIVE

❋

The Tour de France was on. My sister called to say that the racers had passed through her tiny village. To the embarrassment of her husband, she brought an American flag to wave. My sister sometimes wears a tank top that reads, in English, BULLETPROOF BABE.

She wanted to know how the rabbits were. The young bunnies had grown plump and cute and were driving their moms crazy. I finally took them out of the cages and let them run free on the deck, where they would fatten up.

Meanwhile, I was starting to look very thin. One friend used the word "gaunt" and made a sucking noise, drawing her shoulders up and in. It was true, my pants had gotten a little baggier than usual. I cinched my belt buckle on a never-before-used hole. I weighed myself at a pharmacy: 128, my high school weight.

After almost three weeks of vegetables and the odd sprinkling of duck, I was getting hungry for some rabbit meat. I had enjoyed the rabbit in France so much. The meat was tender and light, not heavy like the fatty duck. But I was, as I had initially worried, finding it difficult to kill, clean, and eat a fellow mammal.

Riana's French in-laws—especially the eighty-year-old matriarch of the family—were rooting for me. She had given Riana the skinny on how to kill and clean a bunny. She used a pair of pruners to make a cut in its throat. To clean it, "Mamie says you just pull off its pajamas," Riana reported.

These relatives were gold mines of culinary information. On the last day of my visit to France, I had bought some escargot plates at a flea market. Over dinner that spring night, I asked Chantal, Benji's mom, about cooking snails. While taking a mental inventory for my newly hatched plan, I remembered that I had quite a few snails who lived and bred on my artichoke plants. I wondered if they might add a protein boost. But I had no idea how to cook them.

"You boil them," Chantal said in English with a classic French accent. She looked like a small fox, with reddish hair and large brown eyes, seated at the table. Her hands were graceful and quick and punctuated her instructions. "First you keep them away from food—how do you say?" She looked at Benji. "Starve them. Then cook them in water for an hour." Dress them up by frying them in garlic and butter, she added. Benji said that at that point, the mollusk was usually removed from its shell to have its poop sack excised, then stuffed back into the shell. Sounded easy. (I have yet to get that desperate, though.)

Chantal's parents, Mamie and Grandpa, had been farmers and winemakers, and they had raised rabbits their whole lives. When the Germans occupied France during the war, they took all of Mamie's rabbits, including her breeding stock. The family nearly starved to death. That was why Mamie and Grandpa continued to raise rabbits into their eighties. Bunnies were a symbol of survival.

Although my parents didn't depend on them for survival, their bunnies had been a good source of protein. One night I called my mom to get her thoughts on rabbit killing.

"Boy, did they multiply!" she said. "You have to kill them just to keep their numbers down."

"So how did you actually do it?" I pressed. It's one thing to hear a story about our childhood bunnies and my mom's biology lessons on the farm, but another when I was going to have to execute one. The French could cut a rabbit's throat, but I was sure I would botch that delicate operation.

"Well, I can't believe I could do this," she said, "but I'd bash them in the head with the blunt end of the hatchet, then chop their heads off." And then, I imagined, she would pull off their pajamas.

"I still have the hatchet. Do you want it?" Mom asked. "It's Swiss-made, just needs to be sharpened."

On day twenty-two, the day before I was to do my first bunny kill, the hatchet arrived in the mail. I wondered if doing these things—reenacting my mother's chores on the farm, learning about a process from my elders—would somehow help me understand better where I came from. As I looked at the hatchet, I giggled at the thought that most daughters are given their mother's jewelry or silverware when they reach a certain age. I had received a dull hatchet.

I walked out to the deck and grabbed a white male rabbit from the litter. He squealed when I picked him up by the scruff of his neck, as if he knew his fate. I cradled him against my body, but he still struggled. Once we walked into the garden, the rabbit went limp. I set him on the grass under the plum tree. He sniffed and chewed on the round leaves of a nasturtium. He looked beautiful under the tree, as if he belonged there, as if he was home. His white fur contrasted with the dark leaves of the plum, his haunches resting on the orange flowers of the nasturtium.

I didn't bash his head. I decided to put the hatchet on a windowsill—an art object/family heirloom. Instead, I used a method that I had watched, step by step, on the Internet.

I put the bunny's neck under the wooden handle of a rake. Then I stood on the handle and pulled the rabbit's back legs upward. There was a faint crunching sound as his neck dislocated. Knocked unconscious, the rabbit jolted a little. Then I cut his throat with a sharp pair of pruners. Bright red blood dribbled out onto his fur. I looked him in the eye the whole time, watched his eyes fade and become cloudy and opaque.

Killing Harold had been a Thanksgiving sacrifice, a mercy killing, and a coming-of-age for me as a farmer all rolled into one. This time I just really needed to eat. Though this experiment was a self-inflicted folly, eating a rabbit was going to erase my chronic hunger pangs and give me a few whispers of satiety. That made something that seemed barbaric—killing a cute bunny—very necessary.

I hung the rabbit in the plum tree to bleed. I used a coat hanger and tied the back legs with baling twine to make skinning the rabbit easier. I made a few hesitant cuts with a pair of kitchen shears until the pink flesh under the fur revealed itself. Then the pelt started to peel off, just as Mamie had promised. I had to make only a few more strategic cuts before the whole hide came off, inside out. Underneath was a layer of skin and blood vessels.

As I did the work I whistled and was only slightly paranoid that a neighbor would pass by and try to talk to me, figure out what I was doing, and run screaming from the scene. I was obscured by the plum tree but still felt a little exposed.

Once the fur was removed, it was just a matter of gutting. The entrails spilled out via gravity, making the word "offal" make a great deal of sense— they literally fell out, and they were kind of awful. I couldn't believe how big the stomach was, but I shouldn't have been surprised, as rabbits have several stomachs to digest all that vegetable matter. The bladder was see-through and held a tablespoon of yellow urine. Per the suggestions of the French, I left the kidneys attached, at the back of the carcass. I removed the heart and the liver, which consisted of four dark red lobes, and one small green sac (the gallbladder, which I would later separate out).

I killed and dressed (or undressed, as the case may be) my first bunny in ten minutes. A chicken takes at least thirty minutes, a duck over an hour— another benefit of the rabbit.

The rabbit started to look like those I had seen in the French market. His lines were good—he had plump haunches. To see the flesh that I helped make was a blessing.

I wondered how a vegetarian would have fared on this experiment. She probably wouldn't have been as flip as I had been and would have carefully sown chickpeas and beans. But it's hard to grow enough soybeans to make tofu on a small bit of land. I knew that I couldn't have survived without eating the rabbits I had raised.

After the body was clean of offal, I cut it from the coat hanger, leaving two white rabbit paws hanging in the plum tree. I felt a surge of nostalgia for this moment. Here I was, like a peasant woman, killing my supper, white furry paws hanging among dark plums, me wearing a bright blue apron with a pair of kitchen shears in the front pocket. The mise-en-scène of the tree, the bench below it, a mat of nasturtiums twinkling in the shady spot, the propagation table with trays of small sprouts emerging.

I understood everything about the dinner I would have that next evening, after the meat rested for a day. I had seen the rabbits born, I had carefully fed them fresh greens and snacks from the garden, the Dumpster of Life, and Chinatown. I even knew their personalities. This white rabbit had been the largest of the group—and the bully, always beating up his smaller brothers and sisters—so he made the most sense to kill first.

I hesitated at the entry to the garden, in that boundary between farm and city. Across the street, near the vestibule of the abandoned building, there was a crackhead guy on his hands and knees, searching for something on the ground. He stood up and walked around in circles, examining a dollar bill, ripping off the corners.

I've never been scared of this man. He has never talked to me or even approached me. But today, with the body of the rabbit still warm in my arms, I felt as if I might actually scare him. If he looked over at that moment, if he could think clearly, if he could see what I had done, would he be just as disgusted with me as I was with him? I would explain that I was very hungry and needed comfort. And perhaps he would say the same to me.

Not wanting to scare the downstairs neighbors, I swaddled the slightly bloody carcass in my blue apron and carried it upstairs, thinking about my mom. When my sister and I were children, Mom was making the most of her situation. And wasn't that what I was doing, too? Another restaging of the uniquely American fascination with the agrarian lifestyle. Looking back on my parents' history and comparing it to my present, I recognized that if my parents were Utopia version 8.5 with their hippie farm in Idaho, I was merely Utopia 9.0 with my urban farm in the ghetto, debugged of the isolation problem.

I cheated and used salt to make a brine to draw the blood out of the rabbit. Before I submerged it, I admired the rabbit's lean lines. The saddle—the meaty back section—is the prize cut, and many restaurants serve only that. As hungry as I was, I wouldn't be wasting anything. The back haunches looked well exercised, plump. I put the fur and the head in the freezer for when I got around to learning how to tan the hide using the brain—an old Davy Crockett–era trick. One day I'd make an awesome rabbit-fur-lined sleeping bag. The entrails went into a hole I dug next to a fig tree—they would provide nutrients.

My plan was to invite Bill for dinner. For the past three weeks, I had been eating to survive, mostly grazing while in the garden, so we weren't eating together much. It felt important to be making a whole, hot meal, and I was proud to share the meat of my labors with my beloved Bill. Though he is a scruffy auto mechanic who would be happy to subsist solely on burritos, he happens to have one of the best palates I've encountered. He can sense the presence of secret herbs in fancy restaurant dishes, artfully describe a perfectly ripe peach as if it were a vintage wine.

With the rabbit, there was plenty to share, and so I was doing what a primate hunter does with a big kill: distributing it. A chimp researcher mentioned in *The Primal Feast,* Craig Stanford, noted that "chimps use meat not only for nutrition; they also share it with their allies and withhold it from their rivals. Meat is thus a social, political, and even reproductive tool."

I had been working on the bad-breath thing (flossing maniacally), and I hoped that once I shared some meat with Bill, he would, well, you know, give a little back. Only Bill wasn't a chimp. The meal had to be good if I was going to get any.

The next day, I followed my sister's recipe, which came via Mamie. While I fried the rabbit pieces in duck fat, I thought about my sister. She had her own version of utopia, too. She had gone from an SUV-driving, Botox-using Los Angeles lifestyle to a happy, quiet existence in a rural French village. Perhaps it was her hippie DNA expressing itself, or maybe Mamie's thrifty influence, but in France, Riana had gotten into crafting her own soap and making her own cloth diapers. We were both planted in places wildly different from Idaho, and yet our hidden traits were coming out, adapting to make something new.

Once the pieces of rabbit had turned golden, I poured a bottle of my wine over them. In a 350-degree oven, the meat cooked for an hour with sprigs of thyme and cloves of garlic. I set the table and called Bill to dinner. I served generous portions of the rabbit: two pieces of the saddle for each of us. I spooned a few stewed plums and some sauerkraut next to the rabbit.

We sat down for our first meal together in a long time. The meat was flaky but firm, and redolent of garlic and herbs. Bill took a bite, and I watched him carefully.

"This is better than chicken," he said, smacking his lips and slicing off another piece of juicy meat. Then, be still my heart, he gave me a sloppy kiss before stuffing more rabbit into his mouth.

It was the most flavorful rabbit I had ever eaten. While I chewed, I couldn't help but think of the white rabbit that had been killed so that we could eat. I was thankful that he had been born and thrived on my farm. His flesh became my flesh.

CHAPTER TWENTY-SIX

In the end, they got Bobby.

I wasn't there, but a graffiti artist on our street told me about it.

The city came again and dragged away the numerous abandoned cars and collected the belongings Bobby had built up since the last purge. The man with a clipboard was there again, and when Bobby walked over to him, the police arrived and arrested Bobby.

They told him if he didn't stay off 28th Street, they would put him away.

"We're all doing something illegal on this street," I said to the graffiti guy.

"Shit, yeah."

"I've got all these animals, you're tagging buildings, Lana had that speakeasy, Grandma has her underground restaurant. . . ."

"But Bobby was out in the open," he said.

Someone power-washed the street. A FOR SALE sign went up near where Bobby had once lived.

CHAPTER TWENTY-SEVEN

On day twenty-five of my monthlong experiment, I passed by Grandma's and saw a new sign posted. Another fish dinner. My mouth watered, remembering the tender catfish, the golden cornmeal coating. I had to tell Bill, because someone had to enjoy that food. When I got home, he was in the tub.

"Want a dinner from Grandma?"

He nodded.

"Fish or chicken?"

"Fish."

About ten minutes later, I found myself seated in Grandma's kitchen, looking through her photo album. Her kitchen was small but orderly, with colorful oven mitts decorating the walls and a lethal set of knives above the stove. Cast-iron pans bubbled with oil. I was late, but Grandma was willing to make me up a meal.

I sat breathing in the cooking smells through my pangs of hunger—or not hunger, exactly, but a heady desire to eat something besides a salad or rabbit or apples.

"I'm making meatloaf and potatoes tomorrow," she told me as she bustled around, breading the catfish, then dropping it into the hot oil. "I wish I had some collards."

Could I have imagined her words? I mean, I had hoped that things were going to work out like this, but . . .

"Er—I have a whole bed of collards," I said.

"Did you hear that, Carlos?" Grandma yelled, and brought Carlos into the room. He was tall and skinny and wore a red baseball hat. He had had a stroke, so he was legally blind.

"I have a big bed of collards. I could harvest them for you tomorrow," I offered.

"Good," Carlos muttered.

"I have some Swiss chard. I could bring that, too," I said.

"No. Do. Not. Bring. Swiss chard," Carlos asserted and walked back into the front room.

"He hates chard," Grandma whispered. "Now, I can pay you for the collards," she said.

"No, no," I said. "You just make me one of your dinners. We'll trade."

My dad had told me that hunters often experience a sort of giddy high upon making their kill. As I walked down her stairs, past the flowers and the pebbles, and then past her boys, I experienced a thrill that must be similar to the hunter's high. Tomorrow, I was going to eat, and eat good.

The next morning, after my usual meal of grated young pumpkin and a mug of minted green tea, I went out to the lot and harvested the entire bed of Southern Georgia collards. They had been growing for four months now, but unlike the lettuce, they loved the heat and were still thriving. Also, I had been side-dressing them with rabbit poo. Every week I had cleaned out the rabbits' cages and tossed the black circular turds next to the collard plants. The greens had grown lush and large; their big leaves sucked in the sun and grew larger and larger.

For lunch I had my usual salad, with lettuce, roasted beets, a boiled egg, a bit of leftover rabbit, and an apple. Around two o'clock, I finished soaking and washing the collards and walked them down MLK to Grandma's house. The four grocery bags full I carried would eventually melt down to just one pot of greens. When I handed them to Grandma, she squealed.

Later, I went back to collect my dinner and brought Bill with me.

We sat in Grandma's kitchen and watched her cook as she told us fishing stories. Carlos joined us at the table with a small bottle of Cisco, a red malt liquor, glad that I hadn't brought any Swiss chard. One time, she and

Carlos caught one hundred catfish, Grandma said. They had to put them in the bathtub to be cleaned. Grandma showed us a photo of the fish in the bathtub.

"Hon, you want cake or cobbler?" she asked.

I began to salivate. To eat cake, something made almost entirely out of flour and sugar, a dense load of carbs, well, I would take that. Especially when the cake had pink frosting. Technically, based on my rule number five (Bartering allowed, but only for crops grown by other farmers), I was cheating. Grandma wasn't a farmer. She was a hunter-gatherer. But for me, after almost a month of the 100-yard diet, I started to see boundaries and categories in a different way. Farmers, who had lately become cult figures, were just trying to survive. My snobbery around food had evaporated. Plus, I really wanted some cake.

Bill and I returned home, lugging our almost bursting Styrofoam take-out containers down the dark street.

Then we saw Bobby.

He had relocated one block up. He looked good. He had clean clothes and was wearing shoes.

"They're taking care of me now," he said as we hugged, and he gestured to the big guys in front of Grandma's. I felt bad. Why hadn't I helped him?

Bobby told us he had a new place with a "million-dollar view"—of the highway. He pointed to his spot, and I saw he had built an altar on top of an electrical box to mark it.

"I love you guys!" Bobby yelled.

"We love you!" I yelled. And I meant it. We really missed Bobby.

I lived off Grandma's dinner for three days, like a cougar feasting on a deer. I was happiest about the fish, because I knew Grandma had caught it herself, and I remember the photos of her standing there, proudly holding up her mess of fish. The rest of the food—the homemade potato salad, the mac and cheese, the fluorescently pink cake, a dab of collards from my garden—it was good, too.

CHAPTER TWENTY-EIGHT

❀

Two days before August 1, I saw a newspaper lying in our lot, in the duck area. It was early in the morning, and I didn't have my glasses on. I peered down but didn't see any of the ducks waddling back and forth.

When I walked out to the duck yard, I was greeted by silence. The birds had been massacred. One of the ducks had been ambushed while he slept—he still had his head tucked under his wing. Another had made it—or been carried—over the pen's gate. The geese were dead, too, on opposite sides of the yard, as if they had tried to fend off whatever had killed them but had been divided, finally, and died alone.

It was a beautiful sunny day, still cool in the morning as I stood there looking at the grassy area littered with feathered corpses. White feathers on green—it was strangely peaceful. The animals had only a small puncture or two on their backs.

I kneeled down to touch their bodies in the grass. They were freshly killed, still warm. Peg, our hillbilly neighbor, walked past wheeling a bundle of laundry.

She called in, "Everything OK?"

"No," I said. "They're all dead."

"I saw 'em. A pack of stray dogs run outta here around six this morning," she said.

Stray dogs—this place was really getting third world, apocalyptic.

She continued on to the Laundromat down the street. And I got my pruners and cut off the birds' heads. I wasn't as sentimental as I had been with Maude or the duck and goose killed by the opossum. I let these birds bleed into the ground.

Later I carried the bodies upstairs and hung them from the shower rod by their webbed feet. I put some water on to boil. Even in my despair, I couldn't help but notice how beautiful they all looked, hanging there like bounty from a hunting trip. My bathroom had been transformed into a hunting lodge.

I had never experienced such abundance in all my time on the farm. Two geese and five ducks. What usually is a celebration of a farm animal's life, the melding of its body with mine, became a salvage job. I wondered if it was prudent to eat the meat at all. But I was hungry, and the work of processing them became a meditation.

I plucked until my fingers ached. The goose feathers were particularly small and difficult. Bill helped. He plucked his outside, setting the feathers free to swirl around the neighborhood with the tumbleweaves. I collected mine in a bag, hating to see anything go to waste. I planned to make a pillow or a down vest.

I placed the duck and geese bodies in the fridge. Luckily, there was plenty of room, as it was almost bare. Then I got out my cookbooks.

As I read Elizabeth David, Michael Ruhlman and Brian Polcyn's *Charcuterie,* and *The River Cottage Meat Book,* I realized that the mass killing of my waterfowl, as tragic as it was, had opened up my culinary horizons. Usually with a duck, I would roast it, eat as much meat as possible, then boil the carcass for broth. With the windfall on my hands, I could make some recipes that, because of frugality, I had never dared.

I sliced off the breast meat from a few of the ducks. Following the recipe in *Charcuterie,* I re-created a gamey but delicious duck prosciutto I had had in France. To make it, I simply rubbed the breasts with salt, added a coating of pepper, and stowed them in the fridge for a few weeks. The geese I put in the freezer to make goose sausages at a later date.

Chef Hugh Fearnley-Whittingstall, the author of *The River Cottage Meat Book,* would be my guiding light. He inspired me, going on in his British way about duck confit: "Having a jar just sitting in the larder, bursting with savory potential, makes me salivate every time I see it."

A confit is meat preserved by storing it in a thick layer of rendered fat.

First I had to render the fat. So the next day—after the usual twenty-four hours of rest for the meat—I turned the oven on low. I placed the breastless ducks in various cast-iron pans and cooking trays and set them in the oven. Every hour or so, I poured off the pan drippings. Once the legs were cooked through, I placed them in a jar and poured the rendered fat over them.

I wouldn't say the meat was the best I've ever tasted. Because the fowl hadn't been immediately bled out, it had too much blood in it to serve it to company. But I was a hungry urban farmer, and in saving the meat, I felt almost as if I had saved the birds.

The 100-yard feast (or fast) came to a close. I simultaneously wanted to suck down a cup of coffee and to never let the experiment end. I would miss that slightly hungry, spry feeling. I would miss having my choices limited. I would miss my intimacy with the garden. When I was eating faithfully only from her, I knew all of her secrets. Where the peas were hiding, the best lettuces, the swelling onions.

When I went back to shopping at the supermarket, all those choices would open up again. I could choose from forty-seven different kinds of French cheese. On a whim, I could eat pizza. Or gelato. These are the wonderful things about life—and I made them more precious by not partaking for one short month.

On the eve of July 31, I surveyed the kitchen cupboards. I was down to my last few bottles of plums. I had stripped the last corncob that morning. There were two eggs left. The last honey jar had three fingers of honey remaining. A little bit of sauerkraut marinated in the fridge. A few jars of jam lingered in the pantry. The vat of balsamic vinegar, made from the wine, quietly transformed in the corner. The duck in the confit jar burst with all that savory potential.

In the garden, I looked at the vegetables I would be eating in a few weeks. I anticipated the dry bite of the orange-fleshed, strangely warty Galeuse d'Eysines squash, the crunch of the sweet corn, which still hadn't ripened. The Brandywine tomato that started green in June and turned only slightly pink by the end of July. The forming heads of the crinkly Melissa cabbage. My friends the zucchini plant, the bean vine, the apple tree, which somehow never ran out of fruit.

Hanging in the plum tree, the white rabbit paws danced in the wind. I walked over to the beehive and held my ear to the box. Whistling and rumbling. Who knows what the bees thought of their new home, but they hadn't left.

I had fed myself from my little plot of land, it was true. I had survived, thrived even, through a mix of luck and moxie. But I couldn't have done it alone—I had needed the help of the creatures and plants and people around me.

In the end, the brussels sprouts never formed heads; the stalks were lined instead with some frilly leaves, which I fed to the rabbits and chickens. Toward the end of the summer, though, I harvested cucumbers and tomatoes every day. These I distributed to Peg and Joe. They didn't seem surprised. Peg just said, "Oh, OK," when I handed her a bag of the drop-dead ripe tomatoes and crispy lemon cukes.

I wasn't monogamous with food distribution. We extracted honey in the fall, and I brought Mosed a jar. I foisted tomatoes on every neighbor, every passerby I encountered. The rest I took to the Commemoration Committee for the Black Panther Party and their literacy program. Boxes and boxes of tomatoes. When Melvin Dickson saw me walking through the door with one of the crates, his face broke into a wide grin.

The production of food is a beautiful process. Germination, growth, tending, the harvest—every step a miracle, a dialogue with life. But after the 100-yard diet was over, sharing became the main point for me. I could have hoarded all the food for myself—processed the tomatoes into cans and pickled the cucumbers. I would have had a groaning cupboard of homegrown food. But then I would have eaten alone.

The visitors to the garden kept coming all through the summer and into the fall, sometimes at night. They didn't need to come under the cover of darkness, though—when they plucked a tomato from one of my carefully tended vines, they had my blessing. I understood it more than ever: we were all just trying to survive.

PART III

PIG

CHAPTER TWENTY-NINE

❁

W hat's a barrow?" I asked Bill. He shrugged. We were sitting in some metal bleachers facing a sawdust-filled ring, feeling a little self-conscious. We were 150 miles from Oakland, up north in a town called Boonville. It was April and we somehow found ourselves at a swine auction.

I fiddled with the auction sheet. After almost ten years of beekeeping, vegetable planting, and chicken tending—and more recently turkey raising and rabbit herding—Bill and I were going to reach the pinnacle of urban farming: we were going to raise a pig.

I had discovered the auction while at the feed store buying rabbit pellets. The multicolored flyer, posted on the corkboard amid ads of horses for sale and rototillers for rent, had caught my eye. I stood looking at it as insane thoughts streamed into my mind. By the time I had finished settling up with the clerk for the pellets, I couldn't think of one reason not to go to that auction. I scribbled down the details.

Bill and I had followed the road that led to Boonville, a windy, tree-lined affair, with no idea of what to expect. Upon entering the fairgrounds, we were handed a sheet of paper on which each piglet for sale was listed by breed, followed by a number (which I deduced was weight), followed by a B or a G. Barrow or Gilt. I knew it had to be something about the gender of the pig. Could it translate as easily as Boy or Girl? That the basic terminology eluded me should have drawn my attention to the fact that, this time, we had really gotten in too deep.

Boonville was a sleepy logging town that smelled of Douglas fir forest and wood smoke. It reminded me a little of my hometown of Shelton, Washington. Here, though, if you breathed in deeply enough, you might detect a

fragrant hint of marijuana, the area's number one cash crop. This made for a
tight community, and clearly the other auction attendees knew one another.
There were handshakes and hugs all around. We got vague smiles.

An auctioneer wearing a big black cowboy hat and tight Wrangler jeans
held a microphone and was rattling off numbers on the sidelines. Near the
ring was a ramp where the pigs made their entrance. A large woman wearing a dirty white T-shirt herded a pale pink piglet into the ring. He kicked
up sawdust and squealed.

"That's a lively pig, a good pig, good pig, good pig," the auctioneer yelled.
"One-fifty, one-fifty, do I hear one-fifty?" Then it started to rain—a cold
spring deluge—on the tin roof of the auction barn.

The piglets were the projects of children from local 4-H clubs. The
USDA-sponsored 4-H (Head, Heart, Hands, and Health) program was
started around the beginning of the twentieth century with the idea that
young people would lead the way for more innovative farming. In addition
to encouraging kids to grow plants and animals, 4-H aimed to instill in
them a connection with nature and to promote good old American agrarian
thrift.

The pint-sized 4-H-ers wandered around, helped rustle pigs, and
brushed the piglets before they scampered out into the spotlight for bidding.
The trouble was, there were only about ten bidders sitting in the fairground
bleachers and almost twenty-five weanling pigs.

Bill leaned over and whispered into my ear, "Are you going to bid?"
Nobody was bidding on the pink piglet in the ring. Too high-strung, maybe?
I couldn't tell. The piglets so far all looked the same to me.

"Yeah. Well, I don't know. Which one should I get?" I felt panicked, like
a cornered opossum. The pink piglet went unsold and was prodded out of
the ring.

"Maybe this little one," Bill said, his brown eyes widening. A white piglet
with a black stripe around its belly pranced into the ring.

"OK," I said, and held up my placard—a number 40 in black with red
and white checkers around it.

"We have one-fifty," the auctioneer shouted. "Do we have two hundred? Two hundred?" A jolt of electricity filled my body. I had bid!

A man in a gray sweatshirt sitting behind us waved his number.

"I have two hundred," the auctioneer said. "Do we have two-fifty?"

I had no idea a piglet would cost so much. I shook my head no at the auctioneer. We had heard it was best to buy a piglet from a quality place, so you could make sure its genetics were good and it would produce quality pork. Also, I was into supporting the 4-H kids. But we could have bought a $40 piglet from a farm—we didn't need a prize pig—I just didn't know that then.

"Let's go look at them again," Bill said.

We dashed through the rain to the back of the auction ring, where the pigs waited in pens before they were herded up the wood-shaving-strewn ramp that led to the ring. The pigpens were surprisingly clean, with fresh cedar bedding in each of the twenty stalls. Little children wearing cowboy hats and boots ran around in packs, surveying one another's piglets, which lolled around and napped together.

There were the classic pink ones, deep smoky-red ones—the kids told us they were called Durocs—and Hampshires, black with a white belt across their shoulders. Four tiny red pigs with white faces nestled in one pen by themselves. "Herefords," a lady standing nearby told us. We asked if we could buy two of them outright, but she said they were spoken for. So back to the auction ring and the cold bleachers.

"Here's the one," Bill said, grabbing my knee.

It was a G, whatever that was, and G was cute. It had deep red hair and black hooves. A wheezing swineherd wearing a dirty T-shirt had lofted this piglet into her arms and cradled it like a toddler, with its hind legs almost wrapped around her waist, its head calmly gazing over her shoulder. The woman gave the pig a little kiss before she set it down.

A piglet in a large ring is, by itself, quite wonderful. This little one pranced out into the middle of the circle with a joie de vivre that I could appreciate. The piglet snuffed the ground with its snout, its curly tail flitting back and forth, its hooves kicking up wood shavings. Then suddenly the pig realized that everyone was staring. It let out a shrill scream and bolted for the door.

I suppose I could come up with some lofty reasons for what had gotten me here, to a swine auction in Boonville. To discover the American tradition of pig raising. To test my farmerly resolve in the face of an intelligent, possibly adoring creature like Wilbur in *Charlotte's Web*. To walk in the footsteps of my hippie parents, who had raised a few hogs in their day.

But I'm not going to lie: this was all about pork.

From the moment I first saw the flyer for the swine auction I had thought about all the products of the pig. Smoked pork chops, which Bill and I loved to buy from the Mexi-mart. Pork ribs, slathered in spicy barbecue sauce. Bacon, that temptress, which we preferred cut thick and spiked with pepper. Ground pork, to be used in marinara sauce, or clustered on pizzas, or rolled in sage and fried for breakfast. Sausages, glorious food that feeds the masses, I imagined snuggled up in buns, doused in mustard, and served to all our friends at a barbecue. Ham, of course, smeared in maple syrup and spiked with cloves, was part of my pork daydream. I would be able to make all of these things if I could find a way to raise some pigs. There were other more exotic items I fantasized about as well, like salami and prosciutto. But these were intimidating pork products; I wasn't sure what went into making these, but I knew they were expensive and I liked eating them. I knew that before I got too carried away with my pork-fest fantasies, I had to take the first step: buy a piglet.

Pint-sized pig wranglers waiting on the edge of the ring tugged on the piglet's ear to get it away from the door, and it yelped and moved back to center stage. No one was bidding on this adorable red pig. I couldn't tell if there was anything right or wrong with it. It was kind of small at thirty-eight pounds. The auction was nearly over, and the metal grandstand seats had almost emptied out—there were only three men left, wearing baseball hats and denim jackets, perched like crows in the bleachers. For reasons I couldn't understand, they were sitting out on this pig.

But I wasn't. I enthusiastically waved my number.

"One-fifty. Sold at one-fifty," the auctioneer called. He seemed relieved. The swineherd woman beamed at me. Clearly I had bought her favorite pig.

My first pig. Was it a girl or a boy? Why hadn't I prepared for this? "Excuse me," I asked a towheaded eight-year-old. "Does G mean 'girl'?"

He looked at me as if he might fall over from the sheer power of my enormous idiocy. Then he nodded, so stunned by my stupidity he couldn't speak.

Patting myself on the back, I watched them herd her back into her pen. The little red pig scampered back, as if she knew something good had happened.

A few more piggies were sold, and we neared the end of the list.

"Well, I guess that's it," I said to Bill, signaling that we should go.

"Wait—shouldn't we get another one?" Bill asked. His eyes darted back and forth. The look in his eyes conveyed a deep sense of panic, fueled by scarcity.

"Really?" I said, thinking about our small backyard. How exactly was this going to work?

"Yes, get two. Get two!" he yelled.

One of the 4-H kids had told us earlier that pigs like company, so it did seem like a better idea to get more than one. I didn't want some lonely porker snorting in the backyard.

The last pink pig trotted out. I didn't want it. I had overheard someone in the crowd say that the pink ones can get sunburns. Plus, I wanted a matching set of red porkers. No one bought the pink pig. The towheaded boy standing next to the bleachers wailed into his mom's arms. "They didn't buy it," he sniffled. Poor kid.

A pair of red piglets trotted into the ring.

"We got a brother and a sister here," the auctioneer yelled. "Prize pigs. Make a bid and choose which one you want."

Somehow this confused and attracted me. I held up my number. Sold: $150.

"Do you want the barrow or the gilt?" he yelled over at me.

"The boy," I answered dumbly. The barrow.

After we paid at a card table manned by 4-H parents, we were told to load up our pigs. The parking lot was muddy, and all manner of trucks—Ford

F-250s, Dodge Rams—maneuvered to pick up their pigs. We had a station wagon. Bill eased our car closer to the pigpen and opened up the hatch. An iron cage someone had given us sat in the back, waiting for our piglets.

The owners of the pigs—one of them the woman in the soiled white T-shirt—gave me their phone numbers in case I had questions. "She's a real good girl," the woman said, holding the pig in one arm like a small dog before she put her into our cage.

The 4-H kids helped load the boy pig into the car. He was definitely larger. Once they were both in the cage, they sniffed each other's butts and pressed their snouts together in greeting. Though they weren't from the same litter, it seemed as if they were going to get along.

Near the loading area, someone was selling pig brushes and something called pig chow. It came in a giant dog-food-like bag marked with numbers having to do with ratios of protein and calcium.

Now that I was closer to our pigs, I could see that their hair was fairly substantial, not unlike a man's beard. I figured I should buy a brush. I had visions of morning grooming sessions with the pigs in our backyard. They would stand very still while I coaxed their bristles into a high sheen as the BART train rolled by. A vague thought of buttermilk baths crossed my mind, too.

The woman at the table asked if I wanted a bag of chow to get me started. I shook my head no. We had other plans for feeding our babies. I did buy a bright pink brush, though; it had soft bristles and a sturdy canvas handle.

Nobody asked where I lived or what I was planning to do with these pigs. They just waved as I drove away from the Boonville Fairgrounds. Bill frantically rolled down his window, even though it was chilly out, turned to me, and said, "What's that smell?" I took a deep breath. Oh my god. It was barnyard and sweat, but worst of all a manure odor far too reminiscent of human fecal matter.

Oblivious to their stench, the pigs fell asleep. Bill and I drove in silence, afraid that if we opened our mouths to speak, the pig smell would enter us, land on our taste buds, and go down our throats. We were brined in the choking turd odor for the three-hour journey home.

Back in the ghetto, I herded the pigs into their new home. It was just getting dark; the air felt heavy, but it wasn't raining. That morning, in anticipation of the piglets, I had enhanced the chicken area with a bucket filled with water and a feeding trough made from the same metal washtub I had once used to dip and pluck Harold. I had also tipped a barrel sideways with the idea that they would sleep in it. The chickens had walked into the barrel and pecked at the metal trough; they seemed confused.

Safely behind the closed gate of their yard, the pigs seemed mildly curious but far from geniuses stomping out Morse code with their cloven feet, thank god. They ran around kicking up sawdust, as they had in the ring a few hours ago. The chickens came down from their roost to investigate their new houseguests. The piglets sniffed at the chickens, which caused a panic, and the chickens retreated back to the chicken house for the night.

The pigs were both Red Durocs. Durocs, sometimes called Jersey Reds, are known for quality fat production. These pigs had, even at their young age, the classic arched backs that one often sees in profile on meat-company labels. They had curly tails, but it was not a tight curl, and I noticed later that they wagged them when they were happy with a certain food item, or the sun was shining just right, or I was scratching their backs with a stick. The tails, then, did convey emotion.

As they checked out their surroundings they made quiet grunting sounds. I wondered if the city noises—a police helicopter circling the 'hood, someone yelling at the junkyard dogs next door—would be a shock to their system. They had, after all, lived in the deep country, all trees and pastures. But if they were disturbed by the city's smells and sounds, they made no sign of it. Pigs, I was glad to see, were not very sensitive.

Besides being inspired by pork, Bill and I got the pigs partly out of sheer loneliness. Not only had Lana left and Bobby been removed, but the Nguyens had suddenly abandoned us. Their daughter, Phuong, had packed her car one day and drove to Los Angeles to start college. Their son, Danny, and Danny's wife and baby relocated to an apartment across the street, leaving Mr. and Mrs. Nguyen alone. Accustomed to three generations under one roof, the couple had decided that the idea of staying in a two-bedroom apartment alone did not appeal. I tried not to take it personally.

So they moved to the other 28th Street, the one across the main drag and directly under the highway, to a room in a house with another family. (Meanwhile, Bill and I lived alone in our two-bedroom apartment yet bickered about closet space.) The Nguyens had moved slowly, packing up items and walking them over to the new place.

After two months and nary an apartment-viewing visitor, Bill and I had become very loud. We did laundry at two in the morning, played records at full blast. I clomped around the house in clogs, which I used to take off in deference to the Nguyens. And so when I happened upon the auction ad at the feed store, key in my mind was that there was no one around to mind a backyard swine addition.

A germ of the pig idea had been percolating in my brain ever since the successful Harold experiment. And the bounty we had discovered in the Dumpsters because of the rabbits had also gotten us thinking. A homegrown turkey or rabbit was delicious and made an amazing meal. A homegrown pig, on the other hand, would be delicious and would make hundreds of amazing meals. Homemade sausage, pork chops, all manner of charcuterie, honey hams, and finally, finally, I could face the gateway meat that had turned me from a vegetarian back to a carnivore: bacon.

Would a spider weave messages on her web urging me not to kill the pigs? Were they really as intelligent as everyone said—and would I end up keeping the beasts as seven-hundred-pound pets that would fetch my copy of the *Times* every morning? Or would the pigs turn me into Ma from *Little House in the Big Woods,* with me rendering pig fat and smoking hams all day? Of course it would be the latter—I knew myself well enough by then. A pragmatic farmer, not a soft sentimentalist. Right?

This endeavor was not without risks, however. One big one was flavor. Bill and I were betting that we could feed them from the city waste stream—the bread Dumpster, the Chinatown green bins. But I couldn't find anyone who had actually ever done this. There were no books on the topic. Until we tried it, then, we wouldn't know if it would work or not. Would Dumpster-fed pork taste gross? Would six months of pig husbandry yield undelicious pork? This was like high-stakes poker: heavy losses and heavy wins were both within the realm of possibility.

Bill and I looked at the pigs newly installed in their pen as they nosed around the corners of the area. Then they stood in front of their gate and smiled up at us expectantly. We read their minds: Where's the pig chow? On cue, we jumped in the car and raced over to Chinatown.

That night, for the first time ever, Bill and I threw open the Dumpsters with our hearts—and minds. Will they eat, we wondered, these soggy pieces of Chinese doughnut? I discovered: yes. These chunks of leftover duck from the restaurant window in which everything exudes a steady flow of oil, including this duck head? Yes. Wontons and dumplings covered with, somehow, frosting? Yes. Grapes? Yes. Watermelon? Yes. Egg-fried rice? Yes, yes, yes.

Bill and I anxiously unloaded our two buckets of slop from the car. We had never collected such a disgusting assortment of salty and sweet, meat and vegetable. But pigs, I had heard, were omnivorous, and so we were respecting that.

When we walked through the gate to the backyard, we were greeted by two grunts—one deep, demanding; the other softer, questioning. I hefted a bucketful of Chinatown into the metal washtub trough. The pigs began feeding before the second bucket was empty, so I ended up pouring a load of grapes and wontons over their heads and watching it all bounce off their shoulders and land on the straw-strewn ground.

Their focus was amazing. While they ate, the pigs let out small sighs of approval. Their lip smacking was audible. At times, they would stop chewing and simply suck up the juices from the trough through their nostrils. They were the best dinner guests ever.

The pigs stopped eating for a moment and gazed up at us. Their mouths moved continuously; their chins were smeared with frosting and grease. Now that I thought of it, these pigs had probably never had food like this before. They had probably only had their mother's milk, a few handfuls of pig chow, and maybe a rotten apple. Now they were eating Chinese—like good urban pigs.

The rabbits had always been too finicky to eat any old Dumpster item. Like too-cool teenagers, they looked at me with disbelief when, after a bad night at the 'ster, I would offer them a semi-soggy head of lettuce. One sniff, and they'd hop away. The lettuce would quietly rot, untouched, until I finally scraped it out of their cage. The chickens were only slightly less choosy. But the pigs, I was happy to see, would clearly eat anything.

Bill and I, coated with Dumpster grime, looked at each other in wonder. What had we gotten ourselves into? When the pigs discovered, at the bottom of the trough, the lopsided cake we had dredged from the Yummy House Bakery, they let out peals of delight louder than the squealing brakes of a municipal bus. They bit each other's ears in order to get a bigger share of the cake. I made a mental note for next time: more cake.

Reassured by these eating machines, I knew that—with the help of a pork-motivated boyfriend—it was going to be easy to raise pigs in Oakland. We had seen enough evidence in Chinatown to make our case: All that food could support several pigs. I would soon learn, though, that in this moment of self-satisfaction I was forgetting one key thing: these pigs would grow. As they steadily gained weight they would demand more food than I could ever have dreamed.

That night they wiggled into the barrel together, sleeping head to ass, a drift of wood shavings dusted over their little bodies like a blanket.

A few days after the 4-H auction, my mom called to check on how the pig farming was going.

"I told Dr. Busaca about your pigs," my mom began, "and he said, 'I remember pigs in Oakland.'" She laughed at our family dentist's humor.

"Really? There were little pig farms around here?" I asked, missing the joke. I had been reading that many American cities in the seventeenth and eighteenth centuries had pigs within city limits. They served as living garbage disposals—and sometimes, disgustingly, sewer digesters. In New York City, I read in *Pigs: From Cave to Corn Belt,* the Bowery district had many a pig: "Hog pens projected into the crooked streets. The slaughterhouses were erected astride the ditch outside the wall [of the city], the waste being carried slowly and malodorously down the East River."

Less gruesome sounding was Pig Keeping Council, started after the First World War. According to Michael Hough in *City Form and Natural Process,* "Since the bulk of edible waste came from the cities, pig and poultry keeping naturally evolved as a major urban activity. . . . Pig keeping spread on to bombed sites, in back streets and allotments and included policemen, firemen and factory workers among the devotees." By 1943, Hough reported, there were 4,000 pig clubs, with a total of 110,000 members who kept 105,000 pigs in London.

So I wasn't surprised to hear that there had been pigs in Oakland. Getting excited, I imagined acres of hogs down in the flatlands. I wondered how the pig farmers took care of odors, a new problem I had to solve myself. I became a little dizzy—I was repeating history over and over again.

"Not real pigs," she said. "Cops."

"Oh!" I said, my bubble popping, and laughed at my literal-mindedness. I had forgotten that, as a matter of course, political activists like my mom had regularly referred to the police as pigs. The moniker was also used by the Black Panther Party, whose newspaper often dressed up a cartoon pig in a police uniform. In her book *Framing the Black Panthers,* historian Jane Rhodes describes one of these cartoons, which had the caption "A Pig is an ill-natured beast who has no respect for law and order, a foul traducer who's usually found masquerading as a victim of an unprovoked attack." This cartoon stuck, and people like my mom—and her dentist—used the term liberally.

When I was growing up—and as I learned more about farming—I had been hungry for stories about my mom's time on the ranch in Idaho. I now recognized that I was looking to find my heritage through these rural stories. But the longer I lived in Oakland, the more I wanted to know about my adopted city as well.

Since my mom and dad had both lived in the Bay Area in the 1960s— she as a political-science student at UC Berkeley, my dad as a classical guitar player in Oakland—I would have thought that they would have lots of stories to tell. They even lived together in West Oakland at one point: after they met in Mexico (my mom still insists that it's not a good idea to meet your life partner while on vacation), they shacked up near the Port of Oakland, about twenty blocks from where I live today.

But neither of them could remember much about Oakland back then. My mom had a vague memory of buying tamales from the lady living next door to them, but that was all. My dad said they lived near some Black Panthers and wannabe rock musicians, but when pressed, he couldn't recall much else. Luckily, I had Melvin Dickson.

When I dropped off some lettuce at the Commemoration Committee for the Black Panther Party office one day, I mentioned to Melvin that I was interested in hearing more about Oakland's history.

"We called West Oakland Chocolate City," a rumbling voice called from across the room. Melvin, fumbling with the bag of lettuce, smiled and introduced me to his friend Ali, who sat at a table in the corner of the office. He was short, about sixty years old, and wore a black beret.

"There were black businesses, nightclubs, a major jazz scene down there," Ali said about Oakland's 7th Street in the 1940s and 1950s. The railroad porters—men who cleaned the elaborate, hotel-like train cars and served those who traveled in them—formed one of the country's first black unions. The Black Porter's Union was headquartered in West Oakland, a major nexus for the railroad lines. From those stable, well-paying jobs sprang a community. But it wasn't just African Americans, Ali said; there were Norwegians and Chinese people, too—a multiethnic community in which people mostly got along.

"But they broke it up," Melvin said, sighing.

"They" was the city of Oakland and the federal government and something called urban renewal, Melvin told me. First came the construction of Oakland's main post office, in the heart of the burgeoning black community. Though the post office was supposed to provide jobs, the leveling of homes with tanks, actual military tanks, alienated many. And when the jobs did come, there were only a few.

Then came BART, which used eminent domain to raze hundreds of homes and businesses. To cap off the destruction, they built an expressway and highways 24 and 980 through predominantly African American neighborhoods. Melvin and Ali said this so-called development bisected communities, ruined businesses, and destroyed the close-knit community that had thrived for years. There was no question that these neighborhoods had been slated for destruction because they were the least politically powerful. Later came the crack epidemic of the 1980s.

Melvin and Ali got out a photo history of the Black Panthers and paged through it with me. Here was Lil' Bobby Hutton, killed by the police though he was unarmed. Here was a Black Panther rally, everyone sporting a gun. Violence begetting more violence.

Riding back to my farm in GhostTown, I took Shattuck instead of Martin Luther King, which led to a newly developed corner of North Oakland called Temescal. Several new restaurants had opened up—high-end Mexican, a fancy bakery, a pizzeria with a wood-fired oven. A booming economy in the Bay Area was fueling the revitalization, and new condos were sprouting up here and near downtown Oakland. Art galleries and coffee shops opened their doors, and Oakland's new face—white, professional, artistic—came in. Suddenly, this was the place to be. I liked that I could finally get a decent cup of coffee nearby, but there was something unsettling about all the new development.

When I turned west onto a street that led to Martin Luther King, I rode under the overpass—an uneasy corridor, especially at night, although the acoustics were good for singing—and entered a different world. I passed by Bobby and his homeless encampment along the BART tracks. The bullet-hole-riddled walls of Brother's Market. The shuttered houses that ran along the highway. The decrepit signs for businesses that were no longer. I suddenly saw my neighborhood for what it was: an artifact, an abused landscape. But it could morph again.

CHAPTER THIRTY

✧

On a May night, Bill and I hit rock bottom. At least we hit it together. As a crisp evening breeze blew we stood on a sidewalk in Chinatown, looked both ways, then hurled open the lid to the garbage bin in front of Yick Sun fish market.

A few weeks after we bought the pigs, I found a book—*Small-Scale Pig Raising*—and I had been learning all about them. With the book's help, I had pieced together that we had bought a eunuch and a virgin. A barrow, technically, is a castrated male. A gilt is a female who hasn't had babies. We had purchased shoats (thirty- to fifty-five-pound adolescent pigs), not piglets, as I had been calling them. I had also read that the human alimentary tract and human metabolism are very closely related to those of pigs. That's why people take pig-thyroid medicine and why pig valves are used for human heart transplants.

Besides these little tidbits, I had also read in *Small-Scale Pig Raising* that young pigs need protein. Lots of it. The book mentioned that Norwegian pig farmers long ago fed their pigs fish in order to fulfill their protein requirements. This made the pigs taste fishy, though, so the Norwegians "finished" them for a few weeks on a diet of corn or fruit to remove the fish taste.

Nobody does this anymore, of course. Most pig farmers feed their hogs "rations"—pellets containing a mix of corn and soy. There's a special feed for each stage of a porker's life. Pelleted feed was not on our pigs' menu, and I hoped this would make the pork taste better. But I realized that, as much as they liked our fruit-and-cake diet, it wasn't going to get them the protein they needed.

Fish guts would.

In the center of the Yick Sun garbage can, a black bag quivered with liquid. Bill, as bold as ever, ripped it open. We almost fell down from the fishy blast emitted from this tear. After a few seconds of head ducking and dry heaving, we peered into the bag. Fish heads, guts, scales, tails. I lined up a bucket, and Bill sloshed out a measure of the fish guts. A sickening, chunky stream came out. Some of it splattered onto my glasses, and I yelped.

Just at that moment, a homeless man we sometimes see in Chinatown approached. He appeared to be crying and shaking his head at us. Normally he asks us for change when we encounter him, and we've also seen him sleeping in doorways. I looked up from our focused fish-gut pouring and saw him walking toward us, a crumpled dollar bill in his hand. He couldn't speak, but once he got to us, it was clear: he wanted us to have this money. In the eyes of this man, we had not just hit bottom, where he hung out. In his eyes, we were clearly in much worse shape than he was.

I started to giggle. Bill pushed his hand away. "It's OK, man," he said. We had to refuse the money a few times before he finally shuffled off.

Was it really OK? I wondered on the drive back home, fish guts sloshing in the backseat.

Based on the pigs' reaction, the fish guts were an unqualified success. The squeals of delight were louder than those that any Yummy House Bakery cakes had elicited. The pigs sucked and snorted up the glorious bloody fish guts, chomped on the heads, sampled the mackerel livers, and licked every scale off their trough. Though this was good news for the growing pigs, a shadow of resentment crossed my mind. We'd have to keep going back.

To keep from becoming completely resentful, I had to remind myself again why we were doing this. Pork, glorious pork.

Pork, according to Jane Grigson, the British author of the definitive *Charcuterie and French Pork Cookery,* is a historic meat: "European civilization— and Chinese civilization too—has been founded on the pig." Pigs, I had read, have lived with humans since 4900 BC, in China; some scholars suggest that they were domesticated far earlier than that.

We owe our delicious breeds of porkers to a man named Robert Bakewell. Around 1760, he crossed the plump, short-legged Chinese pig with the long-legged European boar. The result, according to Grigson, was pigs "too fat to walk more than 100 yards." Using these portly pigs, Europeans carried on their centuries-old tradition of making salami and prosciutto and curing lardo, the back fat of the hogs.

In America, pigs arrived with explorers and missionaries. These porkers met the cruder needs of the American pioneers, namely, for salt pork and bacon, preserved forms of pig meat that wouldn't rot and traveled well. Bacon crossed the plains with the pioneers; barrels of salt pork were sent down the Mississippi River.

Our breed of pigs, the Red Duroc, I had read, was created from a cross between the Jersey Red, brought from England in 1832, and the Duroc breed, brought from Portugal in 1850. In *History of the Duroc,* author Robert Jones Evans raves about the Jersey-Duroc breed: "He had within his makeup the characteristics that were bound later to make him a leader in swine production. There were strength of character, ruggedness, prolifacy and the ability to put on pounds of pork on forage and concentrated feeds. The Duroc has been developed through more than three quarters of a century of careful consideration for these qualifications, necessary to make the best machine to convert grain and grass into pounds of pork on foot."

With the knowledge my swine would be big producers, I imagined that I would be able to go either way—fancy, high-end salumi (the broad term for Italian cured meats) like prosciutto and salami—or smoke-cured hillbilly bacons and hams. Despite the trauma of fish-gut harvesting, I was sure it would be worth it.

A few nights later, Bill and I hosted a campfire dinner party. So many of our friends and friends of friends wanted to see the incongruous sight of pigs in our Oakland backyard that we figured it would be prudent to host a meet-and-greet. Kind of like a debutante ball for the pigs. I had found a woodworking shop that would gladly part with as many bags of wood shaving and sawdust as I wanted, so before the party, I sprinkled an extra bag in the pigs' yard to sop up fishy odors and their natural hoggy twang.

It was funny what people brought as gifts for the pigs: cabbages and turnips—what they thought were prototypical pig foods. Bill and I took the Germanic gifts and kept quiet about the fish guts, duck heads, and Yummy House cakes we fed them.

"So do you have a giant freezer?" a tall surfer-carpenter asked me as we stood in the pig area.

It was the first tour of the evening. Five or six pig admirers had gathered outside the gates while the rest of the party sat by a campfire in the lot. It was a cool, clear spring evening, so clear we could see the stars, even in Oakland. The pigs had heard the commotion of the tour, and though it was well past their bedtime, they emerged from their barrel, hoping for a snack. We could see their breath in the night air.

"I'm going to dry-cure most of the meat," I told the surfer-carpenter, and tossed the pigs a cabbage. "You know, prosciutto, salami . . . ," I said, as pompous as a mother planning a Harvard education for her two-year-old.

The smaller pig nosed at the cabbage as if it were a green ball, and the two chased it around the pig yard. All eyes turned to them. In only a few weeks, they had probably doubled in size. Taller and fatter. Their bellies, which were destined to become bacon, were getting a good layer of fat. Their legs were building muscle, and I couldn't help but think of all those hanging prosciutto legs one sees at a good Italian butcher shop. The pigs, seeing there would be no good food, stomped back to bed.

Though I sounded confident, secretly I had no idea how, in fact, any of my plans for the pigs would work. I had zero salumi-making skills. Though I had spent some time inhaling *jamón ibérico* in Spain and snarfing up salami from Armandino Batali's tiny salumi shop in Seattle, I knew that eating was not making.

But my ignorance went even further than the holes in my salumi-making skill set. I also wasn't quite sure how, following the old tradition of "when the nights grow longer and the days colder," I would kill the pigs. Somehow I couldn't imagine executing them with a gun as Carla Emery suggested.

Her book had a bull's-eye diagram drawn over a cartoon of a pig's head and the words "Shoot here." That image had once made me laugh, but now I looked at it with a growing sense of despair.

Henry, a friend who grew up in Cuba, stood off to the side of the pigpen. I sidled up to him.

"Have you ever . . . ?" I asked, pointing to the barrel where the pigs were settling in.

Henry cleared his throat. "Oh, yes," he said. "My uncle raised pigs in his backyard in Havana." Cuba had and continues to have a pretty hard-core urban-agriculture scene; people there regularly raise hogs and chickens in urban settings. In fact, the longer I did my urban-farming thing, the more I learned about the history of this practice all over the world. In the developing world, urban farming is a way of life. Shanghai raises 85 percent of its vegetables within city limits. According to *Alternative Urban Futures,* 28 percent of urban families in Poland engage in agriculture. In Tanzania, the government encourages the cultivation of every piece of land in the city of Dar es Salaam, where residents regularly grow vegetables and raise dairy animals and poultry.

Though most Americans believe in the separation of city and country, there are pockets of urban farming here—notably in Philadelphia, parts of Brooklyn, Detroit, and East Austin. In San Francisco, I had recently heard about a four-acre urban farm that sprouted up next to the highway. Willow's urban farms were multiplying as well. Though most of us were small-time operations with less than an acre of land, added together, we made a considerable-sized farm. My role in this scattered-acres concept was, at least temporarily, the pig part of the idealized, rhizomatic farm. And like many urban farmers, I found myself in over my head.

"How do you kill them in Cuba?" I asked Henry, who has curly black hair and looks a little like Prince.

"We stab them in the heart," he said matter-of-factly.

I nodded, and we went back to the fire to make s'mores.

Stab them in the heart? I thought, looking at Henry in the firelight. That's so romantic. But it might be a little too intimate.

After our guests went home, Bill and I sorted through the rest of the food gifts for the pigs. Some of our friends, in pig ga-ga land, had lost their senses and brought perfectly edible food for humans. A bag of only slightly blemished peaches. Perfectly fine potatoes. We took these upstairs and ate them ourselves.

While I bit into an incredibly ripe peach, my two pig problems floated to the front of my brain. One was how to kill them. The other was, once that was done, how on earth would I process them? I had a little over five months to figure it out.

As the hills began to turn gold for the dry season and Bill and I settled into our twice-a-day pig-feeding routine, Willow came over to our house, her dark curly hair in braids and her car packed with chicken cages. It was late May and we were going on a field trip to Vacaville, a rural town an hour's drive away, to buy some heritage-breed chicks. First, though, Willow had to meet the pigs—she had missed their coming-out party.

"Oh, they're so wonderful!" she said. The pigs were snoozing in a pile of lettuce, the chickens politely scratching and eating near them.

"Yep, they're the big concept," I said, poking a stick through the fence to scratch their backs. "Now I'm a real farmer." At parties lately I some-times had to defend my urban-farmer identity. The term "urban farm" had become part of the popular vernacular, and many people—especially real, rural farmers—took umbrage at it. They were especially annoyed when the self-proclaimed urban farmers had only a few heads of lettuce and a pair of chickens. My definition of "urban farming" involved selling, trading, or giving the products of the farm to someone else. There couldn't just be a producer; there had to be a separate consumer. A real farm also had to involve some kind of livestock.

When strangers at dinner parties questioned the legitimacy of the term "urban farmer," I only had to show them a photo of me scratching the pigs' backs with a rake, the auto shop lurking in the background, and the debate was over. My latest livestock acquisition made me feel complete, whole. Every scrap generated in our kitchen went to the pigs. If an egg had a crack, it went into the slop bucket. Stale bread, moldy fruit, rotten milk—all enjoyed deeply by the pigs. Because of our waste stream, raising pigs in the city made a huge amount of sense. And yet this image of me as Ye Olde Swineherder, while affirming that urban farming in America was a reality, also confirmed something else: I was, indeed, a bit nuts.

Willow and I piled into the car and headed north to the chicken farm. We thought this would be more fun than ordering chicks through the mail. She needed birds for her backyard-garden project.

Willow had recently gone to Caracas, Venezuela, a hotbed of urban agriculture. In addition to several massive urban farms in downtown Caracas, Willow learned about a government-sponsored food-growing program for *los ranchos,* the squatter villages in the hills. Along with encouraging some small-animal husbandry, the government provided people with hydroponic grow tables so they could raise their own vegetables and fruit on the decks of their cinderblock houses.

Willow saw parallels between *los ranchos* and our ghetto, so she developed and fund-raised to create a backyard-garden program. In West Oakland, where Willow worked, there were no grocery stores within a two-mile radius, and nary a corner market that sold produce. If you wanted to eat an apple in West Oakland, you were looking at an hour's journey at the least. It was no surprise, then, that residents ate corner-store food—candy bars, chips, and cookies—instead of fresh produce. Instead of taking a bus to the supermarket for veggies and fruit, Willow proposed that low-income people in Oakland grow and harvest their own food in their backyards.

I had the chance to watch the building of one of the backyard gardens. Willow and crew arrived with supplies—wood, soil, and plants. After some concentrated weeding, the crew placed a wooden raised bed (similar to the ones in our garden), filled it with soil donated from the garden center, and then planted all manner of seedlings (collards, tomatoes, celery) that had been grown in the City Slicker greenhouse. Volunteers would return to show the backyard gardeners how to harvest their bounty and plant new seedlings.

Now Willow wanted to add chickens and eggs to the mix. It would be the same model: the crew would bring in a premade coop and the chickens, and drop off the chicken feed for their clients.

The late-May heat of Vacaville blasted through Willow's windows. I was just glad to get out of town and not think about the pigs for a few hours. They had quickly taken over most of my mental bandwidth.

"Have you heard of the urban farming system they had in Paris?" Willow asked. I had been telling her about my latest foray into the hills for horse manure. The whole squat lot was filled with raised beds—we had reached full capacity. Bill, always curious, brought the pigs some of the manure, and they gleefully chomped it down. Was there anything they wouldn't eat?

"No, what is it?" I said.

"In the middle of Paris in the nineteenth century—right in the middle of the city—huge tracts of public land were devoted to market gardens," Willow said. "They would scoop up the manure from the horse-drawn carriages and use it in these massive urban gardens. They also used cloches to grow stuff during the winter. It provided a huge amount of produce."

This is why I loved hanging out with Willow. Where did she get this stuff? In this case, she had read about the Parisian gardens in her favorite magazine, *Small Farmer's Journal*. I later looked it up—the urban farms were mostly two-acre plots, many in an area of the city known as the Marais. At their height, 1,800 of these little plots grew an annual total of 100,000 tons of vegetables. So much produce, they actually exported the excess vegetables to England, Spain, and Portugal.

"This is it!" Willow said as we pulled into an acorn-strewn driveway.

The chicken lady, a mousy woman with a perm, gave us a tour. Hers wasn't a "real farm," more of a suburban place that had once been rural. The whole house and yard had been turned into a chicken-breeding operation. The chickens were kept in fairly small runs, sequestered, I guessed, to prevent unwanted breed crossing.

A Jersey cow lingered in the backyard, which you could see had once been rolling countryside but now was being covered with town homes. Next to the woman's house was one for turkeys, made out of an old shipping container, and five apricot trees. The turkeys—like the chickens, also heritage breeds, a few Royal Palms, some Bourbon Reds—pecked at the fallen fruit.

"How much do you sell those for?" I said, thinking of Harold and my latest batch of turkeys, which I had ordered through Murray McMurray again.

"Fifty dollars," she said.

"Plucked and cleaned?" I asked.

"Yup."

What a bargain. I knew how much work that was. She should charge twice that.

The chicken lady explained that she was feeding the chickens and turkeys no commercial feed whatsoever. The turkeys had the run of the farm; the chickens were fed with whole wheat soaked and sprouted in water. Willow and I got very excited—we were always looking for new tricks.

"I won't feed them any of that soy feed, so it's the best, healthiest meat and eggs," she said. "I can't advertise on Craigslist, though. Animal liberationists," she whispered. "At first I was worried about you two."

I laughed, but I knew what she was talking about. I had started blogging about my farm, and my sister had warned me about animal liberationists, too. So far, I had gotten only a few angry comments from a woman from the House Rabbit Society who kept rabbits as pets.

A guy, a vegan, had posted to ask, "Why have you chosen to raise and kill animals on your farm rather than perhaps raise animals as a part of a community petting zoo for children or a home for rescued and abused animals? In both the latter cases you could derive the benefits of domesticated animals for your plants without killing them." Who, I had responded, would pay for this elaborate petting zoo? I did like the idea of having people meet their meat, though.

"They're like terrorists," the chicken lady said. Willow and I looked at each other. Now, that was going a little too far.

"I've got two pigs," I said. I wanted to reassure the lady, but at the same time, I had a sneaking suspicion that she had a Bush-Cheney bumper sticker lurking on one of her cars. She was not my people.

"Potbellies?" She smiled, assuming I must be raising pet pigs.

"Durocs."

She nodded her head vaguely. "OK," she said. Even she—a woman with a cow in her backyard and several egg incubators in her carport—thought I was a nutball.

"She feeds them from the Dumpsters in the city!" Willow, my biggest fan, said.

"Uh-huh," the woman said. I could tell she was ready for us to leave. So were we.

Willow and I picked out some of the rare-breed chicks: some Chanteclers (big golden chickens), Langshans (mixed colors with feathered legs), and a handful of Araucanas (the kind who lay pretty Easter-egg-colored eggs). They were older than the ones Murray McMurray usually sent and looked quite healthy.

I snuck a peek into the woman's house while Willow wrote her check. Yep, it was as disorganized as mine. Egg cartons piled up on her counters; feathers wafted to the floor. Now that I was part of the farming club, I had come to the conclusion that farming isn't without its downsides. Like my filthy house, for instance. Between my various real jobs and animal husbandry, there was just no time for cleaning. The floor was dirty from all my tracking in of animal droppings and wood shavings. Hay was strewn all over the front stairs. Sticky beekeeping equipment was piled up around the house. For some reason I thought of Lana when I considered my messy house: it was a sign of a busy, full life.

I had volunteered to house Willow's chicks until they were ready to be adopted by her low-income backyard farmers. When we returned from Vacaville, we set the new chicks loose in the chicken tractor that Willow had built. A chicken tractor is not a poultry-driven farm vehicle, as the name suggests; it's a predator-proof chicken-wire pen that has wheels and can be moved to different parts of the farm. Usually it is used postharvest: You wheel the tractor to a recently harvested area, so the chickens will scratch at the leftover crop, stir up the dirt, and drop their nutrient-rich poop everywhere. I set up the tractor under the plum tree, where the chicks could peck at leaf litter and grass.

Figuring that it was warm enough outside, I put my latest batch of Murray McMurray turkey poults, which I had been keeping in a warm brooder, in the tractor with the new Vacaville chicks. There were four poults—one white, three brownish-red—and they had been in the brooder for several weeks. Suddenly released into the world of sunshine and fresh air, the chicks and turkey poults set to work scratching and eating bugs and grass. I smiled. Soon Willow's chicks would be parsed out all over West Oakland. I felt a little like an animal liberationist.

CHAPTER THIRTY-ONE

The pigs were attracting a fan base. Besides my friends, the kids from the neighborhood had taken to visiting them. I don't know how the kids heard about the piggies, but word must have spread.

"Can we see the pigs?" a group of five ten-year-olds asked me one morning at the gate.

"Sure," I said, and led them back to the sty. I had just fed them their breakfast, and the boy pig, whom I had started calling Big Guy, was bogarting the trough, eating plums as quickly as he could. His neck, I noticed, was starting to get a ripple of fat underneath. The girl pig, whom I had "named" Little Girl, quietly gnawed on a cantaloupe that had rolled into a corner. She stepped on the melon with a dainty hoof to hold it in place while she sucked up cantaloupe juices.

"Wow!!!" The kids cheered, pressing their faces against the gate.

"I never seen a pig before," the oldest kid, named Dante, said. He had light brown skin and braided hair.

"Stinks!" one of them stated.

Dante waved his hand in front of Big Guy. Thinking the kid's arm might be toothsome, the pig sniffed at his fingers, gave a little nibble.

"Hey!! He bit me!" All the kids screamed and ran. But they were laughing. It was summertime, and this is what kids in our neighborhood did all day: wander around looking for something—anything—to do.

"Bill told me you got rabbits," Dante said when I walked back out front.

"That's right," I said. Bill worked at home, so he knew everyone.

"Can we see 'em!! Can we?! Can we?" they all yelled.

"OK, OK," I said, and I took them upstairs to the deck. I felt a little nervous, because I didn't want to get sued or accused of anything, but again, these kids were used to just wandering the streets—their parents couldn't be that worried about them.

Did we have rabbits? We had hella baby rabbits. All three of Nico's bunnies had given birth, so it was a full house indeed. Nico was back from Ireland, but she was working on making a documentary, so she hadn't claimed the rabbits yet.

I handed each kid a baby rabbit. The rabbit kits looked like kittens or puppies. They were tiny, with fur as soft as pussy willows, some spotted, others solid brown. The kids did just what I would have done: They held the soft little things up to their cheeks, snuggled them, and kissed them. A gang of small, scrappy kids from the inner city cuddling with baby rabbits might have been the cutest thing I had ever seen.

"I want one," said Dante.

"Me too, me too!" the others yelled.

"Well, they are a lot of work," I said. "You have to feed them and make sure they have water—"

"OK," Dante said, ready to sign up.

"And you have to ask your mother," I said. I put the kits back with their mama rabbit, who made grunting noises and started licking the babies. Groans went all around.

Their ten-year-old's attention spans maxed out, I sent them off to play. I looked outside a bit later and saw that they were sword-fighting in the garden with some bamboo stakes. My friend Max came over and saw them. "If I were a little kid growing up here," he said, "I'd be over here every day, too."

In a way, I did have a little petting zoo, just as my vegan blogger had suggested.

That night, Bill and I went to Chinatown to get the next day's pig food.

"Do they like bitter melon?" Bill asked.

We were going through the green bins like the professionals we had become. "No. Let's just get the bakery stuff and peaches and greens." I had become the expert, because I tipped the buckets every day and observed what the pigs left to sink into an ever-growing pile of rotting muck in their yard. I expected that this bounty would cause a rat infestation. Oddly, I hadn't seen a rat since the day we brought the pigs home.

I had read that pigs are the best converters of food into meat over time. While rabbits efficiently turn grass into meat, they can grow only so large. A whopping 35 percent of what pigs eat becomes stored fat and meat. They just keep growing. Compare that to 11 percent from sheep or cattle. And, I had read, pigs will eat almost anything, with gusto. This I witnessed on a daily basis.

The next morning, I upended two containers of veggies and fish guts, wontons and fruit, into their trough. After a few weeks of fish guts, the pigs didn't seem quite as excited about the protein slop. Within moments, though, they licked up the last bit and gazed up at me, wanting more.

"Bouf, bouf," Big Guy called out. He stood his ground, his red shoulders curving into a perfect half-circle, his floppy ears almost shrouding his eyes. The pigs were growing at a monstrously fast rate. Little Girl, always more polite, nuzzled the chain-link dog gate that kept them contained in our backyard. I touched her nose; it was slimy but somehow muscular. She gave me a little love nibble.

Not only did they want more food; they wanted better food. And I noticed that they preferred cooked food over raw. Lately, if they didn't like something, they would leave it in the trough to rot in the sun. We were going to have to upgrade.

Driven by the pigs' desires, that evening Bill and I decided to journey with our station wagon, aka the Slop Bucket, into a new frontier: the tony part of Oakland, where high-end restaurants were springing up like mushrooms. We cruised past the nouveau-Mexi place, the wood-fired-pizza joint. The sidewalks were lit up at night, so different from the dim streets of MLK. The fine diners were, true to Oakland, a melting pot of ethnicities and ages. But it was clear, these folks had money.

We cruised past restaurant row and slipped through a gate in the back where all the restaurants shared Dumpsters. We left the car running and scurried like rats to the yawning maw of waste.

"What the hell?" Bill said, lofting an entire clear garbage bag filled with still-warm Spanish rice. Like idiots, we had brought only three buckets. Bill gently laid the bag in the back of the car. I took the plunge and fished out an entire pizza, only slightly burned. In five minutes we had found a squashed key lime pie, a bag of still-warm beans, and a container of old romaine lettuce.

Even though it was approaching midnight, we felt like Vegas gamblers on a winning streak, and we weren't going to quit while we were ahead. We continued on, driving to the bread-filled Dumpsters of Life for a suckle. They were filled to the brim, as usual. A few gutter punks sifted through the Dumpsters. One had climbed in. They saw us and started to scatter.

"We come in peace," I said when we got out of the car. We wandered up to a bin. In order to dissuade Dumpster divers, the bakery had been locking the lids with padlocks. I had figured out that the combination for the locks was the street number of the bakery. But the other scroungers and punks, when they found the locks, promptly cut them. The following night, new locks would appear. These were then broken. Someone scrawled the obvious on the side of the Dumpster in silver pen: THIS IS WAR. It looked like the bakery had finally surrendered; the locks were gone now.

"Looking for something special?" I asked the kid with a mohawk and a nose piercing.

"Cinnamon Twist," he mumbled, tossing whole baguettes and bags of dinner rolls out of the way.

This was the bakery's holy grail, an eggy sweet loaf spiked with cinnamon that, best of all, came secured in a plastic bag.

I grabbed whatever loaves were closest. I dropped a few on the dusty pavement, then picked them back up and threw them in the car. I looked back at the punks, who, watching me, seemed a little repulsed, not knowing we were there for the pigs. I waved as Bill snagged a few stale baguettes for the rabbits, and we were on our way. Though it was late, I wanted to hit one more spot.

I had heard through the grapevine about a fancy Italian place on Fourth Street in Berkeley. Fourth Street is a luxury shopping district, complete with a Sur la Table, an Aveda cosmetics store, and bars that sell $20 glasses of wine. "Whole chickens," my source—a fellow Dumspter diver who mostly scrounged for vegetable oil to power his car—whispered when describing the Dumpster at Eccolo.

The pigs needed protein. Maybe they would prefer roasted free-range chicken carcasses to Chinatown fish guts. I know I would.

We drove, under the cover of night, to Fourth Street. Bill idled the car, and I dashed behind the metal door to the Eccolo Dumpster. The rumor was true: fragrant, whole chicken carcasses. Pawing through the pile, I got so caught up in the bounty—fennel stalks, bread soaked in olive oil—that I didn't notice when the gate creaked open.

"Please explain to me what you are doing," a man in a blue Italian suit demanded. I was caught.

I placed the two chicken carcasses I clutched in my hand into the buckets and slowly turned to face him. The headlamp I used for nighttime foraging shined in his eyes. He raised an arm to ward off the beam, squinting angrily. I fumbled with the headlamp to turn it off.

"Well . . . ," I said, assessing the situation. I could lie, saying that I was a mother of five struggling to make ends meet. Then I decided that the truth is stranger than fiction. "I have two pigs, near downtown Oakland," I panted. "And they're really hungry." Compliments get you everywhere, I knew that: "And this is the best Dumpster in Berkeley."

The man smirked. I could smell his very subtle cologne. I didn't want to think about what I smelled like.

"It's a believable story," he congratulated me, brushing his hands together. "Proceed."

I turned my light on and went back to foraging.

"You know," the manager guy said before shutting the gate, "you should really talk to Chris."

"Who's Chris?" I asked, throwing some carrots into the bucket.

"The owner and chef," the man said as he walked away.

"Why's that?"

"He might take an interest in your pigs," he said mysteriously. He saun-tered back into the restaurant to help sell more $35 pork chops.

Buckets overwhelmingly full, Bill and I drove away. The Eccolo smells wafted around the car. I suddenly felt very hungry.

"What'd that guy say?" Bill asked.

"That I should talk to some guy named Chris."

The beans, the chickens, the bread—that was one fine meal for the pigs. And then some, because we got enough food to feed them for two whole days. If we could stockpile like that more often, we wouldn't have to go Dumpster diving every night. The rabbits and chickens had never been this demanding. We had ventured out only twice a week for them. The pigs were another story. I did the math: assuming 2 meals per day, only about 260 meals left to find.

In June, after almost four months of having no one living in the apart-ment below us—not even one call or visit from the Craigslist ad—our land-lord decided to landscape the backyard.

While the apartment had stood empty—and we had acquired the pigs—our landlords had been in their homeland of Benin. Wilfrid, the hus-band, often went to Africa on business and was a classic hands-off landlord. That is, until this harebrained landscaping project came up.

They had decided that the backyard, which was a hard pan of compacted dirt, was the reason they weren't getting any calls from prospective rent-ers, not because the apartment was deep in a violent and crumbling ghetto. And so they wanted to install a lawn, some ornamental grasses and shrubs. I wasn't sure who they wanted to live below us—some suburbs-loving couple? Not likely.

Our landlords knew I kept chickens and rabbits and the periodic turkey. But I hadn't informed them about the pigs.

A red-haired landscaper was in the backyard when I got home from work. She was prepping the yard for sod. "You've got a little farm going on here!" she said when I rode up on my bike. The pigs grunted when they heard my voice.

"Yeah."

"I planted some honeysuckle," she said, and lowered her voice, "to cover up those barnyard smells."

"Thank you," I said. I had been pouring a bag of wood shavings scrounged from the furniture company into their pen area every day to reduce odors, but there was still a whiff of swine in the air.

"It's great that Wilfrid allows this," she said.

"Well, I guess he just doesn't know," I said.

"He saw the pigs yesterday," she reported.

"Really??" I leaned in, suddenly scared. A hollow fear gripped my heart. What if Wilfrid made me get rid of the pigs? I immediately thought that I would have to have a roasted-pig party. They needed to get bigger if I had any hope of hams and salami. "What did he do?" I asked.

"Well, he was showing me the backyard, and he went to point to the chickens, and one of the pigs came running over, making this loud noise. . . ."

I could guess—Big Guy and his infernal *bouf bouf*.

"And Wilfrid froze in his tracks, blinked twice, and then shook his head."

We both started laughing at that.

Wilfrid was an immigrant from one of the poorest countries in Africa. Benin probably had, like China and Ghana, a robust urban farming scene. Maybe he had no idea how abnormal my urban farm—his urban farm—was. Or maybe it was that I paid the rent on time, didn't deal drugs, and had planted a beautiful vegetable garden next door, thereby notably increasing the value of his property. Some might argue I had been causing a bit of gentrification myself. But the pigs—and their odors—had put a stop to that.

Once the landscaper finished installing the lawn, I let the pigs out from behind their gate. They immediately fell onto the lawn, snorting and rolling in the lush green grass. They loved the suburbs! I snapped some photos of the ridiculous sight of two pigs on the brand-new lawn. A lovely smell wafted over from the honeysuckle that wound through the chain-link fence.

RECIPE FOR PIG'S LOAF

Take 5 loaves of Dumpster bread.

Add water.

Add apples.

Squeeze in some miso.

Stir until loaves are soft.

Don't serve hot—their shrieks of pain are unbearable.

I stood at the stove and cooked for the pigs before I left for a quick trip to New York to meet with some magazine editors. I had finally started to make a living by writing articles, subsidized by work at the biodiesel station. In the last year I had quit my other jobs, at the plant nursery and the bookstore.

I filled nine white buckets with bread, apples, and grains—three for each day I would be gone. Before I caught my early morning flight, I did the rest of my chores. The turkey poults and Willow's chicks had been sleeping in the movable chicken pen. When I went out to feed them, I discovered that one of the little brown turkeys had been crushed by the others. Its body was cold and stiff.

"Oh, no! Oh, no!" I said, and picked him up. Though it was June, the weather was still quite cold at night.

The other birds rustled around, jostling for food, ignoring their fallen friend. The pigs would like to eat this little guy, I thought, staring down at his lifeless, matted-feathered corpse. Then the poult moved—just a slight flutter of the wings—and opened his beak, as if he were gasping.

It takes an hour to get to the airport via BART. I had fifteen minutes left to do the rest of my farm chores, eat breakfast, and finish packing before I had to leave. I grabbed the baby turkey and stuffed him down my sweatshirt. He felt like a cold tamale next to my heart. I zipped up so the turkey's head peeked out, and I breathed great gasps of warm air across his head.

Deciding that it was too cold—and dangerous—at night for the turkeys to live outside, I gathered the rest of the poults into a bucket and carried them upstairs. I created an impromptu brooder out of a wooden box we had found Dumpster diving. The other three turkeys were fine with their new home and didn't seem to miss the chickens. Ice Boy was still not looking so good, though. I threw a few items into my bag, contemplated taking a cab, and finished feeding the rabbits and hogs. With forty-five minutes to get to the airport, I finally set the cold turkey under the light and ran out of the house. He stretched his wing just a bit, and I thought: That turkey's toast.

I barely made my flight. In the small bathroom of the plane, I noticed that I had a small puddle of turkey poo on my shirt. Not an urban-farming high point. All this rushing around to get too much done depressed me. Rushing around was part of city life, but I hated when it interfered with the farm animals. If I had more time, I would have been able to save that turkey poult.

On the train from Newark to New York City, I got a call from Bill.

"There's probably a dead turkey in that box—you should take it out before it starts to rot," I told him, and explained how the turkey had gotten chilled and trampled.

"I don't know what you're talking about," he said. "I just see four turkeys running around our living room, making a lot of noise."

"Beep-beep—beeeeeeep," I called with joy over the phone, earning perplexed looks from my fellow passengers. Bill responded with the same turkey call.

After this near-death experience, I made a silent pledge to move a little bit slower, to take my time. I did live in a city, but the stakes were too high to be careless. When I arrived at my hotel, a summer rainstorm hit. I ran through the rain and felt like a god. I had to resist telling the guy at the desk about my victory over death on a farm in Oakland.

CHAPTER THIRTY-TWO

B$_y$ August, Big Guy and Little Girl had outgrown their barrel and were sleeping outside under the stars. Then one day Big Guy decided he wanted to sleep in the chicken house. He squeezed his increasingly large girth through the chicken-sized door opening, rooted around the straw, and lay down to sleep. Little Girl slept in a nest of sawdust outside the chicken house.

The next morning, Big Guy pried himself halfway out of the chicken-shack door and found himself stuck. I was dumping the slop buckets into the trough when I saw him struggling. I giggled at his problem—it was very Pooh-Bear. But Big Guy wasn't some soft teddy bear. In a panic that Little Girl might get more food than he, he made two tremendous lunges and took the door off its hinges. It crashed to the ground, the safety glass shattering, just as Big Guy wiggled out of the way. Free of the door, he made it to the trough, snorting and biting Little Girl's ears.

The pigs had certainly changed my perspective on calories: I was constantly seeking more of them. If I went out to dinner, I loaded up on the rolls left in the bread basket, the mediocre leftover curry, the too-salty pasta. The pigs wouldn't mind. Bill took to grabbing discarded Big Macs and Chicken McNuggets from a nearby McDonald's garbage can. People stared, it was true, but how could one explain?

Mr. Nguyen, perhaps telepathically sensing the great hunger lurking in our backyard, had come over cradling a pink bag of rice. "Chickens?" he said, and pointed to the backyard. He was going to love this. I motioned for him to follow me back. He admired the new lawn and the shredded bark the landscapers had installed.

When he saw the pigs, he yelled, "Oh, wow!!" I had never seen him so excited. "In my country," he said, "we have a lot of pigs."

I'd heard that even in big cities, many people in Vietnam kept pigs. After I showed him our porkers, every few days I would see Mr. Nguyen crisscross the streets and venture into our backyard with a bulging pink bag of leftover rice or noodles.

Every Sunday night, Bill and I headed for our ritual dives in high-end Oakland and Fourth Street in Berkeley. "I wish we could just drop them off here," Bill said, waving at the Eccolo Dumpster—"them" meaning the pigs—"and come back in two months."

Instead, we jumped out of the car and sorted through the trash. Our shirts got splattered with tomato juices; under the gloves, our wrists were smeared with olive oil; and rotten peach juice coursed down our arms. If we had had time to think about it, we would have realized that we had become these pigs' bitches.

B obby was telling me how to kill a pig: "First we'd make a fire under a big metal thing, like a tub, a barrel."

He'd waved me down as I had ridden my bike by his encampment off 29th Street. He pretty much lived permanently in a green belt that ran along the highway and the BART tracks, although the city would come every few months, clean out his collected items, and chase him off. A day or two later, Bobby would return and simply start over again. I'd admired the stuffed animals that lined the fence in front of his place, and before long our conversation had turned to hog killing in Arkansas.

He continued: "Then someone would shoot the hog in the back of the head, then they'd stick it—with a knife or something sharp, get all the blood out." Bobby stood next to the door of the fence. The grade of his steep encampment was almost thirty degrees—he pitched a tent up against a wild plum tree. Stacks of collectibles littered the area.

"But how did you get the bristles and hair off?" I asked.

"OK, OK." Bobby put his gnarled hands up to slow down my questions.

"Then we'd take the hot water, and we'd pour a little onto the pig's skin and pull off the hair. Then do another spot, then another, until it was all pulled out."

"You didn't dip the whole pig into a big vat?" I asked.

"Too heavy—how would you do that?" He crinkled his eyes at me.

Bobby hadn't talked to my mother, who in her pioneer days used a backhoe to loft the dead hog upward and dip him into a fifty-five-gallon barrel of almost boiling water. But my parents had the advantage of living next door—about five miles up a windy road—to some professional hog farmers. The Spelt family hosted an annual pig slaughter, and the first year my parents had pigs, they were invited to bring theirs and share the chores.

"The women were all inside cooking," my mom said over the phone, dismissively, "so they couldn't believe it when I went outside and helped the men." She wanted to see the action, help and learn, not stir the beans. Other than the backhoe, their method was the same as Bobby's: shoot, stick, dip, scrape.

Having this ritual taking place in my backyard, especially in light of its new suburban look, was becoming increasingly hard to imagine. And considering that my key advisor was a grizzled homeless man, my original charcuterie concept seemed impossible. Salami and prosciutto? I might have to settle for some hacked-up pig meat.

When I turned to the library for assistance on making charcuterie, I realized that, in this case, a book was not the proper way to go about learning. The duck prosciutto I had made on the advice of a book had turned out to be edible, but it definitely wasn't like the stuff I had tasted in France. It's not as if I could learn a two-thousand-year-old skill by following some diagrams.

I found the answer to my conundrum, like most things that summer, at the Eccolo Dumpster. As I was rummaging again in its bounty, a young sous chef emerged and wanted to talk about the pigs. He had heard about us from the suit-wearing manager.

I told him about the porkers' powerful hunger, the job of feeding them, and my vague plans about processing them.

"You should talk to Christopher Lee," he said.

"So I've heard."

"He's here all the time. Best to come in the morning," he said before disappearing back into the kitchen.

The next day, still deluding myself that I would be preparing four hundred pork sausages in my kitchen and stringing salamis from my fireplace come September, I bought a meat grinder at Sur la Table.

Eccolo happened to be right next door. It was a beautiful August day, sunny and clear. The San Francisco Bay, just a few blocks away, sent sea breezes down the tony shopping center's streets. Eccolo's patrons sat on the restaurant's sunny patio, eating salads and dipping bread into olive oil.

Fuck it, I thought, and I pulled open the front door and walked into the hushed dining room. It smelled of almond-wood smoke and braised meats. The restaurant was fancy. It seated fewer than fifty. The linen was tasteful, as were the fresh-flower arrangements. Near a big wooden bar was an open kitchen that allowed patrons a view of the chefs at work. All I knew about the place was that it was a descendant of Chez Panisse, the original California-cuisine restaurant.

I was wearing clothes that I considered nice, but compared to the people at the tables, I was basically wearing pajamas. I also noticed that my shirt-front was splattered with beet juice from a salad I had made for lunch.

A blond hostess wearing all black greeted me: "One for lunch?"

"Um, no," I stammered. "I was told that I should talk to Chris. . . ."

"About?" She smiled.

"I have two pigs!" I blurted out.

"OK." She looked a little flustered and disappeared into the kitchen.

It was around 3 p.m. A few people were sitting in the dining area sipping predinner cocktails. I checked the bottoms of my shoes, which were waffled with the fragrant muck of the pig yard.

Though I felt vulnerable and slightly embarrassed, the fact that I had been in Eccolo's Dumpster somehow gave me a mental advantage. A chef's secrets—and disgraces—revealed themselves in the Dumpster. I had seen the stock-soaked bay leaves, the woody cardoon stems, the herb-roasted-chicken carcasses. One night in their Dumpster, I had found two huge hunks of meat that had been rejected for some reason. I carried them home like trophies and cooked them in a vat of boiling water with Dumpstered cabbage. The pigs loved it. They didn't mind an extra dose of protein, or being cannibals. I had witnessed—and salvaged—a chef's failure. It was as if I had gone through Chris Lee's underwear drawer. And now we would meet.

Then he was before me, tall with a head of snow-white hair as soft as goose down. He looked to be around fifty-five. He had kind eyes but seemed slightly annoyed. I hadn't prepared for what I would say. I didn't even know what my goal was in talking to Chris.

"The other night," I started, "I was in your Dumpster and one of your chefs said I should talk to you."

"Oh?" he said.

"Yes, well, I've got these two great big pigs in downtown Oakland," I said, opening my arms to show just how big they were, "and I've been feeding them with your restaurant scraps, so I thought we should meet."

"Why are you doing that?" he asked, squinting his eyes. "Keeping pigs in downtown Oakland?"

He had me there. I had all kinds of reasons. Because I'm an ecofreak, because of bacon, because I can't bear to see food wasted. In a way, the answer was: because I could. I told him about the urban waste stream that I was tapping regularly. The bread Dumpsters, the high-end Mexican, the Chinatown bins. He seemed impressed.

"I don't like wasting food, either," Chris said. "I kept chickens for years and would bring them scraps from the restaurant."

"And I want to feel close to my food," I said, "to see what it means to raise it—and kill it."

He nodded his head.

"In France, I learned of a humane way to kill a chicken," Chris said. "With a sharp knife, they would reach into the chicken's mouth, just under the tongue, and cut an artery." His large brown eyes went a little bleary. "You hold them, and they just go limp in your arms and twitch a little."

"It's solemn," I observed, thinking of Harold.

"Yes," he said, and we both touched our hearts.

I didn't know what else to say.

"When you have a whole pig," he said, "you get the complete picture."

I nodded as if I understood. I had come only to confess to Dumpster diving in his compost bin but found myself blurting out, "Teach me how to make salumi, how to make hams. I'll give you one of the legs from the biggest pig to make proscuitto." It was a Rumpelstiltskin kind of deal, but it was still in my favor.

"Maybe I can help you." Chris looked skeptical. Despite my clear insanity, he and I arranged a tentative plan. He made salami on Mondays and Tuesdays. It's a two-day process, and I would be allowed to watch.

Before I left, Chris gently shook my hand. Like an idiot, I said, "Your Dumpster is just wonderful, Chris. Thank you. Keep it up!"

"We have it cleaned once a month," he assured me, and disappeared back into the kitchen. Off I went, stained shirt and all, visions of salamis dancing in my head.

Back at the farm, I heaved a bucket of slop to the pigs. I was gripped by fear: What if they don't taste good? I fretted. A chef like Chris would have a sensitive palate. Would he recognize a Chinatown fish funk? I scratched Big Guy's head and gently tugged on Little Girl's big floppy ears. They grunted their approval of the bucket of peaches with loud smacking sounds.

It was month four of the pork project, and being the pigs' bitch wasn't so fun anymore. I was exhausted. They were eating four full-sized buckets of food a day. Foraging for them had turned into a part-time job. Every Sunday, Wednesday, and Friday, Bill and I trolled the Dumpsters, desperately seeking enough food for their maws. On those nights, we would fill the car with a dozen buckets full of slop, unload them into the pig-feed staging area in the squat lot, and dole them out until it was time to dive again.

The idea of buying the pigs food never crossed my mind. Well, once— one day I asked the lady at the store where we bought our cat food if she sold pig chow. "No one has ever asked me that," she muttered, and then went back to her invoicing project.

I took the meat grinder upstairs and thought about grinding Little Girl up. It seemed rather horrific. Thank god Chris Lee, a wise elder, would guide me through this, with the most respect possible.

A couple of days later, I ran into Chris while in his Dumpster.

"Can I help you?" he asked from behind the gate in his most withering tone.

"Chris, it's me, it's Novella," I called, and came around to show my face.

"Oh."

"I just scored at the cheese shop," I reported. In our desperation, we had branched out even further and started hitting delis, grocery stores, and a cheese store. I had been relieved at their bounty.

"Oh?" He seemed hesitant.

"Huge chunks of Brie," I said, and just to remind him: "For the pigs."

He smiled. "Oh, OK. So see you next week?"

"Yes, yes, the sixteenth."

Later I e-mailed Chris to apologize for scaring him at the Dumpster. "I admit that my behavior is deviant," I wrote, "but I get so enthusiastic about it, I forget it might freak someone out." He wrote back:

> Novella,
> I'll be ready to go in the morning. Pork arrives Monday. I usually start by 10 am and spread the work over two or three days.
> It was funny the other night—I hesitantly approached the suspicious guy [Bill] in the dirty Mercedes (suspicious/Mercedes oxymoron?) and suddenly you appear from the darkened dumpster area in your miner's hat. I've heard about your farm and want to come see it.

I had the green light. It was without fear that I continued to collect the roast-chicken carcasses, the olive-oil-soaked bread, and the scraps of fine meat that were all part of the glory of the Eccolo Dumpster. My pigs were going to be amazing.

A few days later, I heard a commotion outside. Our new downstairs neighbors' dogs were barking, then I heard a distinct grunt. Bill was soaking in the tub, and we were just chatting about the day's events. I ran downstairs.

I glanced at the gates: they were all open, including the main pig door. As I rounded the corner I saw one of the monks, Chao, in his full burgundy robes, holding an orange parking cone. He was speaking pig. *"Uh-uh-uh-huh,"* he grunted. The pigs were capering around 28th Street.

When they cavort, things break. Big Guy plowed into a garbage can, and it toppled. Little Girl was trying to get into the garden by ramming into another gate with her tremendous shoulders. A bull in a china shop would have been remarkably calm compared to these two. Another neighbor, Sandra, who had recently remarked that my farm reminded her of her childhood home in Puerto Rico, lofted a broomstick, with her daughter assisting.

This wasn't the first time this had happened. A few weeks earlier, I had heard someone outside say, "Is that a pig?" and then I saw Big Guy headed for the busy intersection of Martin Luther King and 27th Street. (Little Girl, being nice, was in the pen, eating like a pig princess should.) I trailed him, yelling and pleading, but he had a date, apparently, or a bus to catch, and he merely glanced sideways at me, grunted, and trotted faster. Even trapped within a well of fear, I couldn't help but enjoy the clicking sound of his hooves on the city sidewalk, his ears flopping in the wind.

As cute as this pig outing was, if Big Guy had made it to 27th Street, he would have turned into inedible bacon, not to mention that I would have had some pretty steep body-shop fees to pay. He weighed about 175 pounds.

Luckily, I spotted some of our 28th Street neighbors up ahead, returning from the corner store. "Can you stop the pig?" I asked.

"What?" a young man with cornrows said.

"Just, uh, you know, scare him," I suggested. The man got in front of the pig. Big Guy stopped.

The man clapped his hands. I could see Big Guy's tiny brain working on how to solve this problem, and whether it was worth all the hassle—and exercise. With a soft *bouf,* he turned his curly tail and trotted back to the 2-8. Me chasing behind, yelling encouraging things.

The guy with the cornrows wanted to process what had happened. As I shut the gate behind Big Guy he asked with a smile, "Do you think

God made pigs to eat? Cuz when I saw that pig, I thought, yum, ham and bacon."

"I don't know," I said, "but I think the same thing. We bred them to be like this, so maybe they become what we want."

Cornrows grinned. "Well, I've never seen a pig before. That was really something."

Feeling like an experienced pig wrangler after that experience, I walked outside feeling relatively calm about the two pigs now cavorting in the street. I went first to Little Girl, who was trying to break down the fence to the garden, and yanked on her ear. She sat down. Tugging two hundred pounds by the ear is remarkably ineffective. "Can you hit her butt with the stick?" I called to Sandra. She came over and gave Little Girl a short smack. The pig budged, a little.

Sandra's daughter and Chao built an impromptu herding area out of fallen garbage cans so that the pigs could be trapped at the end of our dead-end street. The monk worked on Big Guy, who finally got sick of the man in the robes making the grunting noises and went into the pen without anyone touching him. I broke out a bag of bread and lured Little Girl with it—she followed me into the gate.

"Thanks, you guys!" I said from behind the gate, sweating and feeling frazzled by the escape. These people were so nice.

The monk pointed to the schoolyard field to which the pigs had beelined. "They want to be free," he said.

I had to agree.

The next week, on a sunny August morning, I arrived at the restaurant for the first day of my salumi apprenticeship. I made sure to wear a new white shirt and clean shoes.

Chris greeted me, drinking a milky cappuccino—did I want one? I nodded, and a woman behind the bar pulled me a frothy cap. I met Chris's wife, who was arranging flowers at the bar. "Here to steal Chris's secrets?" she asked, smiling. I shrugged. Before we started, Chris told me his one ground rule: I must never share his recipes with another person. These salami recipes represented years of training and fine-tuning. I nodded. All that tradition felt a little weighty, as if I were being indoctrinated into a secret society.

We went to the back kitchen to set up. He had pulled out several raw pork shoulders and a large slab of back fat. We were going to make salami. When Chris bustled out to get a hotel pan, one of the prep cooks, a skinny twenty-year-old, asked me how I had heard about Eccolo and Chris's salumi-making skills. "I Dumpster dive here," I began to tell him. He shrieked with laughter and high-fived me.

Maybe I've read too much Anthony Bourdain, but I had imagined that the back of a restaurant would be a crude, uncivilized place. I expected to get groped, not high-fived. Everyone who passed through this kitchen seemed intelligent and kind.

Samin, the sous-chef and Chris's right-hand woman, arrived and began to slice turnips. She looked about twenty-six and had thick, dark hair. "What are you doing?" I asked.

"Making pickled turnips," she said. She told me that everything served at the restaurant was homemade. The mustard, stewed tomatoes, mayonnaise, pickles, sauerkraut, walnut liquor, and obviously the salumi, which includes all the cured meats, like salami, coppa, prosciutto. She was what she called the grandma of the place.

"I'll show you my pantry sometime," Samin promised, adding that she sometimes made spicy pickled vegetables in the style of her Iranian grandmother.

No one seemed to think it was odd that a Dumpster-diving urban pig farmer was in their midst. In fact, I came to learn that the restaurant industry was filled with other obsessive freaks like Samin, who would never buy a factory-made pickle. I was just another one of the freaks.

Chris came back and put the meat on ice, then we took a quick tour of the cold rooms where the salumi were aged. The whole ceiling was filled with hanging meats, hundreds of salumi stalactites in various shapes and stages of decay.

"We make a finocchino salami, made with fennel," Chris said, and grasped the end of a mold-dusted salami. "Here's a soria, made with hot peppers. Coppa—" He pointed at smaller, more circular parcels of meat. Though it was cold, I could still smell the place—a wonderful combination of mushroom and meat.

"What's that?" I asked, pointing at a gigantic salami about three times the size of the others.

Chris smiled. "That's my special one. I call it the Petit Jesus," he said. "I have to secure it with a net, see?" I could see a raised checked pattern on it—like it was wearing fishnet stockings. The Petit Jesus, he told me, hangs in supplication for many months before it is ready. A waiter came in, consulted the metal tag hanging from a coppa, and then took the meat away to be cut paper thin and served on a large wooden plank decorated with olives. I stood gazing up at all the meat in awe. How many pigs did it represent?

Chris and I walked back into the prep room, and I noticed two prosciuttos—two salted pig legs—hanging near the doorway. They looked exactly like my pigs' legs, but dry and hairless. "Watch your head on these," Chris said, pointing but not looking at them.

Without ceremony, he began to trim a pork shoulder. He set the hunk of meat on a cutting board and cut it in one-inch chunks. These he tossed into the hotel pan (which was sitting on ice), separating them by degree of fat: extremely fatty in one corner, no fat in another. Each shoulder, with back fat added, would make about ten salamis, Chris said.

Salumi are never cooked, he explained as he trimmed the pork shoulder. The meat instead is "cooked" by salt, a bacterial enzyme, nitrates, and time. "So a raw-food person could eat salami?" I asked.

"But we won't let them," said Chris, smiling. Focused on his work, his head bowed over the project, his careful hands deconstructing the shoulder, he rattled off the names of the main muscle groups that made up the pork shoulder and what jobs they'd done for the pig. He told me that his pork came from Oregon and had grazed on hazelnuts.

After weighing and dicing up some pure white back fat, Chris sent the meat and fat through an industrial meat grinder. He had me weigh out the spices and curing agents that went into each kind of salami. We used a digital scale, and everything was measured in grams. I felt as if I was back in one of my college science labs.

We added the spices to the ground meat. "Look in here," Chris said. A whirling mixer was whipping the meat and spices together. It looked like some sort of ghastly cookie dough. "It'll reach a perfect point—there!—when it starts to hold together. This is very important," Chris said, exhaling, and turned off the machine.

Samin peeked her head into the room. "Your son is here," she said. And then a skinny punk-rock teenager slouched in. He was prep-cooking at the restaurant for the summer. Leaving father and son alone for a moment, I ran out to the kitchen to get the salami stuffer, a giant red metal machine. I passed by the line cooks, who were sweaty and concentrating on making pasta and roasting chickens.

I had never worked in a real kitchen before. In college I was a dishwasher for a few months, but the place was just a Mexican joint. This kitchen, in contrast, was amazing. There was an entire wall of spices in mason jars. The pastry chef, a wholesome Chinese American woman, had her own realm, with a Sub-Zero freezer and all manner of sauces within easy reach. A long wooden counter that stretched between the line cooks and the servers held vases of fresh herbs, wooden bowls of eggs, and a million utensils. There were three walk-in refrigerators.

Back in the meat room, I met Chris's son, who also wanted to watch the salami-making. I handed Chris the stuffer, an old-fashioned Italian contraption, which he placed on the counter.

"This is a beef middle," Chris said, showing us the beef intestine we would stuff the meat into. It had been soaking in water and looked like a large condom made of skin. It looked even more like a condom when Chris attached it to the salami stuffer, glopped a load of the mixed-up meat into the hopper, and began to turn the crank. The intestine grew plump with meat. It was mesmerizing.

"This is when lots of dirty jokes get made," Chris said. Acting unimpressed, his son wandered off to prep lettuce for the salads.

After the casing had stretched to about three feet long, Chris popped the meat onto a metal tray and tied the end with string. "There you go," he said after all the salamis had been stuffed and tied.

After we painted a different batch of salamis with penicillin culture, we had a little extra time. Chris said we should make a coppa. He trimmed one of the shoulders into a perfect heart of meat about the size of a football. This we rubbed with smoked paprika and some other fiery spices. Chris brought out a veiny cow intestine, called the beef bottom, and we slipped the heart of meat inside. Then we went around to the cook line and dipped the coppa into boiling water.

"Smell it," Chris said. I leaned in—it smelled like a barnyard.

"I like it." I grinned.

"You would," he said, like we were old friends.

"Let's try one," he said, and brought out an example of what the finished product would look like after four months of aging. The coppa had gone from football to softball size, a white mold had formed around the whole thing, and when Chris cut into it, the meat was a deep red, as if it had been cooked. He handed me a piece. This is food that honors the pig, I thought as I chewed and the subtle flavors filled my mouth. It wasn't like the turkey or a rabbit—merely a delicious and sacred thing to eat; this pig, through alchemy, had been transformed into something higher, almost immortal.

Chris stared at a slice, then we both chewed thoughtfully. It was so good, smoky and rich, earthy. "So that's my deal," said Chris.

In his book *About Looking,* John Berger wrote, "A peasant becomes fond of his pig and is glad to salt away its pork. What is significant, and is so difficult for the urban stranger to understand, is that the two statements in that sentence are connected by an 'and' and not by a 'but.'" I felt well on my way to peasantdom. But I needed Chris to teach me more—and I secretly hoped he would help me when it came time for my pigs to meet their maker.

Every week throughout August, I returned to the restaurant to learn more: How to make pancetta, which is the pork belly rubbed with spices, rolled, and tied. How to make the Petit Jesus, Chris's specialty salami, modeled after the Spanish soriano—large chunks of spicy pork and coarse herbs.

Over that time, as I learned about salumi, I also learned about Chris. He grew up in Illinois, where he learned to cook, but he went west in the early 1980s when the whole California-cuisine scene started to happen. Specifically, he went to work for Chez Panisse, the world-famous restaurant in Berkeley. During some of his almost twenty years there, Chris was a forager—a person who goes out to local farms to find the freshest, most delicious ingredients possible.

Chris was ten years younger than my parents, on the tail end of the hippie generation. Like my mom and dad, who built their own house and raised their own food, Chris had the urge to make a connection to something tangible, something real. One day, as we trimmed some meat, he told me that he had considered starting a farm, too, but in the end decided that he had too much to learn from the city. His craft would be cooking. When he discovered the art of curing meat, it became his lifelong obsession.

It was the 1980s, and a law in America had banned the import of bone-in prosciutto, so Chris began working on making his own for his customers at Chez Panisse. "My first prosciuttos were too salty," he said. "They tasted too meaty, not flavorful like the ones from Italy." Then he got his hands on a small booklet of guidelines for Italian prosciutto. ("There were no books then, as there are now," said Chris.) Though it wasn't a recipe, it did help him figure out that the pig legs he was using were too small. He had to source bigger pigs than his regular supplier provided. He found a farmer in Oregon who raised pigs fed on pasture. The booklet specified that the animals must weigh more than 240 pounds, so Chris asked the farmer to grow them to full size. Also, the booklet said there should be only 3 percent salt—Chris had been using too much.

His experiment was a painfully slow process: it would be eighteen months before he would know whether it had been successful. He remembered at one point looking at tiny flies on the prosciutto hanging in the Chez Panisse curing room that he had built. He fretted about them. But when he went on a tour of prosciutto makers in Italy—where the restaurant had sent him to learn traditional methods from the salumi masters—he realized that the flies were a good omen.

"I remember standing in this world-class prosciutto maker's drying room—it was just massive, just massive space—when I saw the same little flies," said Chris. "I didn't want to be rude, pointing out the flies, so I waited." When the tour was over, he pulled the guy aside and asked about them. "I said, 'What are these?'" The ham maker explained that the flies were a normal part of the curing process, nothing to worry about. Chris was exhilarated.

Next came salami making. Again, through trial and error, trips to Europe, and an apprenticeship with a master salami maker, Chris finally solidified his process and recipes.

Chris said that when some Italian customers found out the salumi plate was made in house, they were "surprised, then dubious, then surprised again. I had one guy say, 'This is very good, but this is not prosciutto.'" Chris laughed. Hard-core traditionalists say you can't make prosciutto in a place so close to the sea. They say it must be a hundred kilometers or more from the ocean, which Chris thought was totally arbitrary.

He cut into the soriano—the Petit Jesus—and held it up to the light. The slice was almost four inches in diameter, with clear distinction between the larger chunks of meat and the finer ground meat. Zingy and meaty, with a spicy start from the smoked paprika and a hot finish from the bird's-eye pepper, this was clearly his favorite. Nobody else in America made this salami.

Chris said if you cut into most salami, you'll see the silver skin, the air pockets, the too-uniform, hot-dog-like look that marks a mass-produced product. This one made by Chris was like a stained-glass window—fat alternating with meat in an inconsistent, artisanal way. It was truly beautiful. And delicious.

Near the end of August, and the conclusion of my apprenticeship, Chris agreed that I could bring one of the pigs to the restaurant after the kill. We clearly amused each other. During my training, Chris had shown me knife moves and I had told farm stories and made jokes. He said it would take two days to deconstruct Big Guy and make, with my clumsy help, salami, coppas, and prosciutto. All I had to do was find someone to kill the pigs.

"There's an Anthony Bourdain quote I love," Chris told me as he trimmed some pork bellies, evening them up so he could show me how to roll them into pancetta, another thing we would be making with my pig: "Every time I pick up the phone, something dies."

"Yeah, I gotta find someone to execute those fuckers," I said.

As the time grew closer, my attempts to hire an assassin were getting desperate. No one wanted to come to Oakland to kill my pigs. One traveling butcher laughed. "We do farm kills," he said, "but you don't have a farm." Maybe he sensed, over the phone, my hackles rising, because he started to backpedal. "I mean, you have a farm, but nobody's going to go all the way down there."

After I hung up, I went out to the pigsty and poured a fresh bed of sawdust. The pigs loved it and rolled around happily. I had to dash out of the pen quickly, though, because lately they had started to scare me a little bit. Little Girl would pull urgently on my shirttails—not because she wanted to tell me something but because, I think, she wanted to drag me down and eat me.

Every other farm animal had taken a backseat to the pigs. I didn't know how many rabbits we had or whether the chickens were getting enough food—I could only think of the pigs. Even my family and friends had taken second place behind the pigs. I hadn't talked to my mom or sister in weeks. Willow went Dumpster diving with Bill and me sometimes, but I could tell her heart wasn't in it. Bill and I had grown closer, though. After a sweaty night of summer Dumpster diving, we would share the bathtub, wash each other's backs, and marvel at the treasures we had found that night, and the porcine treasures that were growing larger and larger in our backyard.

The pigs had grown so big, the other animals were scared of them. The chickens avoided them. I kept the young turkeys, suddenly fairly large, in a separate pen so they wouldn't be eaten.

And though I wanted to kill the pigs myself—perhaps with Bobby's assistance—I recognized that this might be a job for a professional. To harvest a thirty-pound turkey, to kill a three-pound rabbit—in retrospect, that seemed so easy. Done at home, with simple tools. But to kill two animals that weigh more than I do—for some reason, this fact was significant—made it a big deal. Finishing the pigs was to be the pinnacle of my urban-farming experiences. And the nights were growing longer and colder. It was getting to be hog-killing time.

CHAPTER THIRTY-THREE

A fter a few more days of phone calls and e-mails, I found a killer, and a woman at that.

A butcher named Jeff, whom I had found up north, had agreed to break down the smaller pig and directed me to a slaughterhouse close to his shop. Sheila's slaughterhouse. I dialed her immediately. I needed to get a firm date so I could tell Chris when Big Guy would arrive.

"Oh, I hate the Bay Area," she told me when she heard that I lived in Oakland. But she agreed—all I had to do was get the pigs to her ranch, and she would take care of everything.

"I do kills on Fridays," she said, "so just bring them by on a Friday." Any Friday. It seemed rather cavalier. I told her I would be there on a Friday in September, and felt reassured that all was well.

As the date approached I would remember unanswered questions— Where was her slaughterhouse exactly? Should I starve them the day before? Could I get the offal?—but when I called, I would get her answering machine. The outgoing message was in Spanish and English. I left long messages with my questions but received exactly zero calls back, which made me nervous.

Her butcher friend, Jeff, was a bit easier to contact. He had three phone numbers and always answered. He promised that the pig would be butchered to spec, wrapped, and ready to pick up the following week.

"But I want the heads," I said. Chris had told me about the most amazing book, *The Whole Beast*. The author, Fergus Henderson, pointed out that all parts of an animal should be eaten, not just the prime cuts. Eating livers and kidneys and brains was a tasty way to not waste an animal's life. The book begins with "Seven Things I Should Mention." Henderson's number one: "This is a celebration of cuts of meat, innards, and extremities that are more often forgotten or discarded in today's kitchen; it would seem disingenuous to the animal not to make the most of the whole beast: there is a set of delights, textural and flavorsome, which lie beyond the fillet."

I was very eager to try all kinds of delights—head cheese, pig's ears, trotters.

"Oh, you'll have to talk to Sheila about that," Jeff said.

"And the blood," I added, remembering boudin noir, French blood sausage.

Sheila.

But she wasn't answering. On Labor Day, after leaving three messages in as many days, I again called the Wild Rose, her slaughterhouse.

"Novella, we're barbecuing," Sheila said. "We're open Monday through Friday."

"But I need to have a few questions answered," I said, shocked that she had picked up and struggling to remember what my questions were.

"How long does it take?" I asked, finally, lamely.

"Half an hour for each pig," she said.

"And can I watch?" I said, rummaging through my bag.

"Yes. OK, Novella, we'll see you on Friday," Sheila said, and hung up the phone.

Listening to the dial tone, I found my list:

· *what time to arrive?*
· *does she cool them off?*
· *what about the head?*
· *should I bring buckets for the blood and offal?*
· *directions*

Oh, well.

The Saturday before I would take the pigs to get killed, my neighbor—the husband of the beautiful Vietnamese woman who had handed Harold to me across the fence—approached me as I took out the trash.

"Excuse me," he said. We'd mostly just waved at each other, never had a conversation. "Your pigs"—he pointed behind the gates where the pigs were biting each other and squealing over some toothsome bucket of slop—"are smelling very bad."

I nodded. I had cleaned out the bunny cages and thrown the soiled straw into the pig pen, thinking that the pigs would enjoy it. Surprisingly, they didn't touch it. Big mistake. The ammonia smell from the rabbit pee festered in the sun and blended with six months' worth of pig shit to cause a reeking smell. "My little girl," he said, and gestured at a little munchkin wearing pigtails, "almost vomited the other day, it smelled so bad. Can you move them over to the garden?"

"I'm really sorry," I said. "I'm going to have them out of here next week. I'll make sure to put down more sawdust. I'll get on it right away," I stammered.

I bent down to the little girl.

"I'm sorry, sweetie," I said. She lit up with a smile and squeaked. I've never felt like such a complete ass. Who would ever want me for a neighbor? All of this looks good on paper and sounds good in theory, but what about the difficulties and challenges?

I had discovered a fascinating how-to book called *The Integral Urban House,* published in the 1970s, about a house that featured a rabbitry, chicken coops, fish ponds, and composting toilets. The house sprouted up during the eco-movement of the time in an attempt to illustrate a perfect system for growing your own food in the city. I loved the line drawings of a mustachioed man butchering a rabbit.

Chris remembered the house, specifically how the residents of the house used to shock visitors by killing a rabbit or a chicken in order to show them where their meat came from. In the end, the house lasted only a few years—people who lived there couldn't keep up with the work, some of the systems didn't function very well (I heard the composting toilet was a disaster), and of course the neighbors, as they have throughout history, complained.

This was one of my biggest problems. In the case of the pigs, I had to admit it: I had become a neighborhood pest. Then I did what every person who has ever felt that they were in the doghouse, that they were incontrovertibly wrong, that if they were to appear in front of Judge Judy, she would take the other guy's side, would do: I bought a dozen roses—peach ones—from the farmer's market and left them on my neighbor's porch with an apologetic note and an offer of pork chops.

The next day, I saw the man again.

"Thank you for the roses." He smiled and looked at me in a way that told me all was forgiven. But he didn't want any pork chops, thank you. He wasn't vegetarian; I think he just didn't believe that these urban pigs would taste very good.

I was pretty worried about that, too. But I couldn't dwell on my flavor fears, because over the next few days my primary worry was the logistics of the killing operation. I would have to beg, borrow, and possibly steal in order to get the pigs up north. I had no truck. I had no trailer. I considered renting a U-Haul van, but at $200, it was too expensive. At one point, in the middle of my trailer worries, I again contemplated doing the job myself.

"We could borrow a gun from someone in the neighborhood," said Bill. He pointed out that we could hang them from the stairs in order to gut them. It would be easier, I thought. But no, the idea of butchering a pig by reading a book seemed more insane than driving sixty miles north to have the animal taken care of by a professional.

"I'll learn for next time," I said. Sheila, though she had been hard to reach, would surely guide me through the process. She would maybe even let me pull the trigger. It was terribly important that I knew their death would be comfortable and fast, humane and not scary.

I started a countdown for the pigs and posted photos on my blog. Miraculously, no PETA people showed up to free the pigs. We Dumpster dived for the last time on a Wednesday. The pig-meat gods dripped fat upon us, because it was the best haul we'd ever gotten. Peaches in boxes, so easy to load up. As many melons as we wanted. Crunchy lettuces, plums. In the cheese Dumpster we struck gold: spent ricotta, balls of mozzarella—a whole bucket's worth. Bill and I actually had to rein ourselves in. We only needed enough pig food for one more day.

The pigs' noses were coated with white ricotta by midnight, then washed clean by the moisture from the lettuces, and finally sticky with peach nectar. Then they had no more food, except for a bucket of peaches I reserved to lure them into the trailer.

The trailer. I borrowed a friend's Ford F-250 with a trailer hitch and rented a Harley-Davidson trailer from some people off the Internet.

"We weren't going to do it," the guy told me when we hooked up the trailer. "We had some nasty animal experiences. One lady used a trailer to transport her rabbits, and we could not get the stench out."

"Don't worry," I said. "I'm going to line this with a tarp and sawdust and hay."

He nodded. He wore a faded, handmade shirt that read RICK JAMES'S BITCH.

But underneath my calm veneer, I was sweating it. Rick James's Bitch was renting me a trailer that usually transported motorcycles. It was shiny and black. I pulled away, tooted the horn. Back tomorrow afternoon. I thought about how it would all be over by then. What relief I would feel.

Early the next morning, Bill and I plotted the pig capture. We had a series of gates that they would pass through. One held the pigs in; when we had first gotten them, we had even had to add a lock to prevent them from escaping. Another separated the street from the house with a narrow pathway. And behind that was yet another gate. It was the perfect hog run. We parked the trailer butted up to the last gate. I put a bucket of peaches in the rear.

"Load 'em up!" Bill said. He had to go to the auto-parts store to get some glow plugs for a car he had been working on.

"OK." I walked over to their gate. They were groaning with hunger. They hadn't had a bite in something like twelve hours. A record. I undid the gate and—nothing. All the sawdust and food scraps had buried the bottom of the door and wedged it shut. The pigs pushed and nuzzled the chain-link fence. Big Guy crashed all of his three hundred pounds against the gate, but it still wouldn't open.

I got out the shovel and began the archaeology project of unearthing the corner of the gate. As I dug further the lovely smell of ancient mudflat drifted up. Bill walked over, the other gates open, to assist. That's when Little Girl, with all her might, that deep hunger giving her super-pig strength, pushed the gate open, knocking me over. Big Guy followed her. I threw down the shovel and yelled to Bill, "Run!" He took off in front of the pigs, sprinting down the concrete pathway. I'd never seen him run so fast.

The pigs, sensing a fun game, chased Bill. Their pig hooves clattered against the sidewalk as their tremendous buttocks thundered along our walkway—past the garbage cans, the turned-over ladder, up into the trailer. Just like that—that easy. Little Girl paused for a minute, but I prodded her butt and she shrugged and got in. Big Guy jumped into the nest I had constructed in the back of the rig and began feasting on the peaches. We shut and latched the heavy door, then put on a padlock as extra insurance.

I felt like a real farmer with my trailer stuffed with two real hogs. Bill rode off on his bike, and I pulled away from the homestead, the fence to the garden newly tagged with fresh graffiti, the chickens watching me from behind the chain-link fence. The truck pulled against the weight of the pigs in the trailer. Here I come, country!

The drive was unremarkable. To pass time and soothe my nerves. I did calculations of how much I had spent on the pigs: $300 to buy them, $150 to kill them, $80 to rent the trailer, $60 in fuel. Almost $600 in all. Like most things, what that tab didn't include was the time we had spent Dumpster diving and taking care of them. Toward the end, between the two of them, they were eating seven fifteen-gallon bucketfuls every night. Also not included was the time I spent worrying about pig smells and pig escapes.

Organic, hormone-free pork, the last time I had looked, cost about $5 a pound. We were going to get about three hundred to four hundred pounds of pork, and much of that would be turned into even higher value by virtue of curing and preserving it. Almost $2,000 worth of pork, say. Not bad.

I followed the directions to the slaughterhouse over country roads, past stacks of straw and hay and dried-up hillsides. It was Indian summer in the Bay Area—still hot during the day, a little cold at night, incredibly dry because it hadn't rained in months.

I pulled up to the slaughterhouse, where three guys sat at a picnic table drinking Cokes.

"Sheila here?" I asked.

"In there." They pointed.

I walked into a small room that I recognized as the kill room, where my pigs would meet their end. It was concrete with little gates where the pigs got locked in. It smelled a little bloody with some hints of bleach. A man wearing rubber boots sprayed a hose.

I walked outside and saw a woman walking away.

"Sheila?!" I yelled.

"What do you want!" the woman spun around and yelled. "I've got a ranch to run here." Maybe she thought I was an animal liberationist.

"I brought two pigs to get killed," I shouted. "I'm Novella," I said lamely as she walked toward me.

I had been imagining a country woman, wearing her hair in braids and a calico shirt with jeans and boots. Like a butch Laura Ingalls Wilder.

But Sheila was about four-foot-ten, with a tremendous geyser of blond hair caught up in—was that really? yes—a banana clip. She puffed on a slim cigarette, and her quick hands were adorned with a bevy of gold-and-diamond rings. A smear of lipstick traced along her lips. This was not a butch Laura Ingalls Wilder. She in fact resembled a prostitute.

I wore a T-shirt that read CARPE DIEM. I was sweat-soaked from the drive—and the stress of driving a trailer packed with two animals in which I had invested a lot of time.

"I'm Novella from Oakland," I repeated, feeling very country and pastoral and of a place.

"Oh god, you're the girl with all those questions," Sheila said. "I thought, I'm going to kill her, with all her questions."

I laughed nervously, caught off guard. She had never called me back to answer any of my questions, which seemed dangerous. But I had thought that, once I arrived in this woman's place, she would take me in hand and help me. I thought it was like making travel arrangements in a foreign country—it's really best to simply show up.

"OK, bring your trailer up to there," she said, and pointed at a green fence with two white posts that looked to be three feet apart. "And don't hit the post."

I would now have to do something that I had thus far avoided: back up the trailer.

There's some upside-down physics to doing it, some steering-a-boat logic that I simply cannot comprehend. If I turned the wheel to the direction opposite I wanted to go in, the trailer would go that way. After five attempts at backing up, then pulling forward, I somehow got the trailer through the gates.

The Latino guys watched silently, sipping their Cokes.

I opened the trailer doors. While driving, I imagined the pigs in the back, swaying along with the curves, endlessly jostled and bumped by every canker in the California highway system. But there they were, curled up together, deeply asleep in the bed of burlap and straw.

"Hi, guys," I said.

Shelia arrived on her four-wheeler, platinum hair blowing in the wind.

"Get the pigs out," she said.

"They're just starting to wake up," I said.

"Get in there and get them out," she ordered.

I jumped in the trailer and pushed them up and out. I felt like a clumsy kid, a newbie 4-H-er. The pigs grunted a little in protest, then unfolded and clambered out of the trailer.

"Sorry," I said. "I haven't done this before."

"Well, at least you got them here, that's something," she said.

Then she saw the pigs. "Oh, no, they're red pigs," she said.

"Yes," I said, astonished that she hadn't thought to talk about anything before I brought these pigs to her. It had been one of those "millions of questions" (question number 156,478: Do you scald and scrape the hair from the pigs?) that Shelia never answered when I called her.

The bristle machine was down, she told me, and wouldn't be fixed until tomorrow.

Sensing what a lovely person Sheila was, perhaps, the pigs followed her into their new quarters: a spacious concrete pen. A pink pig snorted nearby. I was glad to note that my pigs were bigger than hers.

While we filled out the "paperwork," a slip of pink paper with my name and phone number, Shelia told me she used to run a beauty store—that explained the nails. "This is what can happen," she said, and showed me her thumbnail—a green shriveled nail bed. Since quitting the beauty store, she'd been ranching now for twenty-two years.

"Sheila, I really want to see the pigs die," I said. "I need to be there. So I'll come up tomorrow morning and watch the whole thing," I said, feeling a little tearful.

Sheila didn't seem able to focus on anything, though, and vaguely nodded. When I started to ask her more questions, she brushed me off, jumped on her four-wheeler, and roared away.

My last view of the pigs: They were in the concrete stall, looking optimistic, sniffing a trough on the floor. They smelled other pigs, and perhaps they thought this was a real farm visit. Like an exchange program for city pigs.

"Bye, guys," I said, and they just looked at me out of the corner of their eyes.

I pulled away from the Wild Rose feeling vaguely uneasy but also vastly relieved. I had successfully raised the pigs without poisoning them, having them escape (permanently), or—my worst fear—seeing them get mangled on I-80. And now I had successfully unloaded them at the slaughterhouse. But the slaughterer didn't seem to be taking me very seriously. And why should she? A city slicker with two pigs—big deal.

For me, though, the pigs were a twice-a-day (at least) interaction, and I had wanted their death to be as respectful as possible. Sheila was not going to be a good killer, I could tell. But I drove away because I had no other choice. I had rented the trailer and delivered the pigs, and she was my only hope.

My suspicions deepened when, at the local feed store, I stopped to get twelve bales of straw (how often do I get the chance to buy a trailer full of straw for the chickens and the garden?). Talk turned to pork.

"Oh, you are going to be shocked how good it tastes," he said. "It's nothing like that stuff you buy at the store."

I mentioned that I had fed the pigs fruit.

"That's going to taste so good," he encouraged. "It might even be a little sweet." I thought I detected a little drool coming out of his mouth.

"Where're you getting them done?" the guy asked.

"Sheila's," I duly answered.

"Oh, yes, she's a real one-of-a-kind character," he said.

"She seems insane," I said, desolate, kicking my legs on a bale of straw. The worker threw tightly bound straw into the trailer. I thought I heard him mutter in agreement.

That night, while at a poetry reading for a friend in the city, I got a voice mail from Sheila: "Novella, your pigs are ready," she singsonged. "I need to get them off the box tomorrow, so call me." Left at 8:10 p.m.

When I called her, she said she had to call me back—some piece of equipment had broken. "OK," I said, hung up the phone, and in the middle of the bookstore I shouted, "Cunt!"

They were dead. And I hadn't been there. Not that I was looking forward to seeing them killed, but I wanted to be there as a way to close the door on what had turned into a massive task—feeding and caring for the pigs. I had also wanted to make sure they hadn't been scared in their last moments. I had hoped that maybe having me there would have made their death easier. People were staring at me in the bookstore, so I wandered outside. I hated her attitude, too. I had told her that I wanted to watch, to help, to be a part of these pigs' death as fully as I had been involved in their life, but that hadn't registered with her.

After I let that go, I worried about the other details. The blood, which

I had hoped to collect to make blood sausage. The heads. The intestines! I dialed Sheila again. No answer.

An hour later, she called. Now I was going to get some questions answered. "I'm covered in muck," she started off, "but I wanted to let you know your pigs are ready."

"Where are you keeping them?" I asked, unsure whether I could even trust this woman to have the common sense to put the pig carcasses in a cool place.

"They're in the walk-in," she said in her best placating tone.

"Now, Sheila," I said, "I told you I wanted to be there. I'm going to let that go. But I want the intestines and the heads."

"OK," she said. "We'll get the heads. And the insides, we have those in a bag for you."

"What about the blood?" I asked, knowing this would be impossible.

"Yuck," she said. "No, we don't keep that. Novella"—she said my name with a Spanish accent—"I have to go take a shower. Can I answer all these questions later?"

I had to face facts: I had entrusted my pigs to a bitch from hell.

"Well, if we had talked about any of this even for ten minutes," I said, madly pacing between people trickling out of the poetry reading, "we wouldn't be having these problems right now."

"OK, so we'll see you tomorrow, honey. Drive safe."

One time in Seattle, while I was riding my bike home from work, uphill, a stranger started to follow me on foot. He grabbed at me, and I had to ride faster to get away. Then he ran faster. I pedaled fast and finally got away from him, but an hour later, I found myself in my car, cruising the streets for the man. I had a piece of lumber in the backseat. I can't explain what I was planning to do, but now I had that same feeling of rage. Directed at this '80s nightmare turned pig murderer.

It was the same feeling of cosmic unfairness that I had felt when the ducks were slaughtered by the dogs, or by the opossum. Injustice. Life gone bad. Gracelessness. And worst of all: It was my fault. Again. Because I had been slightly desperate, because I didn't have very much experience, because I had lost control.

In the end, I wanted to blame America. This is how we do everything: we rush around because time is money, even at the folksy slaughterhouse. And the tradition of not using everything—of throwing all that good stuff away just to deliver me the muscle meat on a hook—made me feel sick. The fact that I was culpable in this fiasco made it suck even more.

Swirling with rage, I drove up with Bill the next day to pick up the carcasses. They were cut in half and hung by their back legs from hooks connected to a pulley system. I couldn't even look at Sheila, I was so pissed off. I wrestled the pig halves off the hooks, and laid them on some burlap sacks in the back of the same station wagon that had ferried them from Mendocino to our home as shoats. Their bodies lay in supplication, stretched out and hairless. I scattered bags of ice over them.

The pigs were pale and cold. Their skin was intact. Since they were cut in half, I could see that both of them had two inches of glorious fat all over their bodies. Big Guy was going to make some terrific prosciutto—what a great big ass. I carefully put their heads—which I was glad to see had expressions of hope on them—into buckets of ice I had brought. Sheila handed me a grocery bag—about five pounds of offal. I glanced inside: a gloppy combo of dark liver and some greenish stuff. It took everything in my power not to pour the contents over Sheila's frizzy head.

We drove away, then stopped and got some peaches at a roadside stand. The pigs already smelled like *jamón ibérico* in the car. I fretted about why I was still so enraged. Here I was, I had the pigs, they looked great, I could relax after six months of hard work. Why wasn't I celebrating? Why couldn't I let it go that I'd missed their death and most of their organs?

With the juice of a September peach dribbling down my chin and the man I loved so deeply beside me, I sifted through my thoughts of anger. Around exit 56 for Vacaville, it suddenly became clear: I had spent almost half a year devoted to making these pigs happy, to truly knowing them, yet in the most important moment of their lives (to a farmer, at least), I hadn't been able to bear witness. Margaret Visser writes in *The Rituals of Dinner,* "When the meal includes meat and especially if the animal is 'known' to us, death can be dramatic. In order to affect people, such a death must be witnessed by them, and not suffered out of sight as we now arrange matters; attention is deliberately drawn, by means of ritual and ceremony, to the performance of killing. This is what is meant by 'sacrifice,' literally, the 'making sacred' of an animal consumed for dinner." Sheila hadn't allowed me to make my pigs sacred, and that was why I was so angry.

I dropped Little Girl off at the butcher shop near Sheila's and made plans to pick up the wrapped meat later in the week. Big Guy I took to the restaurant.

When Chris saw the pig, he said, "Nice pig." His brown eyes ran along the haunches of Big Guy with respect.

His admiration snapped me out of my rage. We had a lot of high-quality pig meat to process. I could mourn later. For now, my days would be all about curing, preserving, and eating meat.

The deconstruction of the pig, Italian style, began.

The difference between English- and Italian-style butchering is one of finesse versus sheer strength. Once we had wrestled Big Guy onto the cutting table, Chris ran his hand over the pig's back. With a small knife, he pressed into the haunches and made a few graceful incisions. The back leg came off easily. I watched in awe-filled happiness. Meanwhile, my American butcher was firing up the band saw.

Suddenly, Chris held what looked like a regular store-bought hacksaw. He crouched into position and began lightly sawing the shoulder and front leg. This came off, and with the side that was left, Chris expertly disassembled them down to recognizable cuts—a hunk of ribs, a rack of loin chops, a chunk of back fat, which I knew we'd need to make salami, a slab of pork belly about two inches thick.

Chris paused every now and then to hit his knife with a sharpening steel. I had the sensation of watching a dance performance. A light shone on him as he performed, a white-haired meat maestro, dancing around my fallen pig.

He picked up a section of the loin chops. "This," he said, "is the most expensive piece of meat." He ran his blade along a squishy pink part of the meat above the loin. Chris sliced his boning knife in and extracted the tenderloin, a three-foot-long tube-shaped piece of meat. "It's tender because it isn't used much. But it's overrated," he said.

On that note, I told him that Sheila hadn't saved most of the offal. There would be no weird lung terrines, no boudin noir.

Chris seemed almost as devastated as I had been. He grasped the butcher table, as if to steady himself.

"It does look like there was a little trauma here," Chris said, and pointed a knife blade at a reddish spot on the muscle part of the shoulder.

"Oh god," I said, and told him about what had happened—how I had missed the actual death of the pigs.

He just shook his head in sympathy. Then, as if to make up for it, he put the pig's back leg on the table, trimmed off a bit of skin, and told me to rub it with salt. We would turn this into prosciutto.

"Is this big enough?" I asked. I remembered Chris's saying the Parma pigs in Italy grew to be seven hundred pounds.

"It could be a little bit bigger, but it will do," he said.

Samin laughed when she came in and Chris told her that I had a complex about the size of my pigs. One of the prep cooks poked his head in and, when he saw my pig leg, said, "That's huge!" I beamed.

"What do Italians do when they're doing this?" I asked Chris as I sprinkled salt onto the leg.

"They talk about women," he said dryly.

I cackled. I could see why. I was essentially massaging the pig's butt. I ran my hands down the skin, pressing, pressing. I finessed the salt into crevices and folds in the skin. The salt rubbed against my skin like sand. After a few minutes of rubbing, the salt drew out water, and my hands were wet. This, Chris told me, was how you knew you were done. I did the same with the other leg.

Then we pressed the pig's legs into a large plastic box filled with salt and weighted them down with a chunk of wood.

It was the same recipe that Cato the Elder recommended in his treatise on Roman farming, *De re rustica:*

> *Take a half peck of ground Roman salt for each ham. Cover the bottom of the jar or tub with salt and put in a ham, skin down. Cover the whole with salt and put another ham on top, and cover this in the same manner. . . . When the hams have been in salt five days, take them all out with the salt and repack them, putting those which were on top at the bottom. . . . After the twelfth day remove the hams finally, brush off the salt and hang them for two days in the wind.*

When I shimmied the big box out to the walk-in, I hit my head on the prosciuttos hanging in the breezeway. These had been curing for eighteen months, as would mine.

"All prosciutto is is salt, meat, and time. T-I-M-E," Chris said, washing his hands, when I came back in.

I told him that I had read a recipe for prosciutto from a time before Christ. He nodded. "Doing this work, it connects us to the past. The past is just a river we all stand in."

While rooting around the history of prosciutto making, I had stumbled upon this quote from Pliny the Elder, the ancient Roman naturalist, about Epicurus, the famous Greek hedonist: "That connoisseur in the enjoyment of life of ease was the first to lay out a garden at Athens; up to his time it had never been thought of to dwell in the country in the middle of town." The garden, as far as scholars can sort out, grew fruits and vegetables. That an urban farmer existed before Christ made me feel like I was—that we all were—merely repeating the same motions that all humans had gone through, that nothing was truly new. This insight gave me a sense of peace.

The breakdown of Big Guy took Chris only two hours. But we still had a lot of work to do: the bellies to roll into pancetta, the salamis to make. Over the next week, I returned to the restaurant to do to my pig everything I had learned during my salumi apprenticeship.

That night I had my first taste of our pork, or pig meat, as Bill liked to call it.

I rolled the tenderloin in a light dusting of salt and pepper, then seared it in a hot cast-iron skillet until it was brown on the outside. I served it with a dab of Tuscan pepper jelly that Samin had made.

When Bill came in from working on the car, his knuckles covered with grease, I took a piece of the meat, swabbed it with pepper jelly, and stuck it into his mouth, then did the same for myself. The flesh tasted extremely sweet, plummy. Tender it certainly was; it didn't even require a knife. While we chewed, the juices of the meat threatened to overflow our mouths. Not overrated at all. Our months of labor had been well worth it. I was relieved that I couldn't detect any fishy taste.

Although Bill had been telling people we would never have another pig again, now he said with a titch of paranoia, "Do we have enough meat?"

The next day, Chris and Samin boned the shoulders. Each of them had one of Big Guy's shoulders, and they seemed to be racing. Both of them had spent time studying with a butcher in Tuscany, so they had been trained by a master.

Normally I would have been trying to find a way to be useful and help, but in their kitchen, I recognized that my role was just the curious, grateful farmer.

"How did you know what to feed them?" Samin wanted to know. She makes most of the calls to their pork suppliers, and after asking what breed they were raising, she always inquired about their diets. The famous Parma hogs are traditionally fed whey from cheese making; in Spain the pigs browse on fallen acorns.

"I had a book," I said. "But mostly I thought about what I would like to eat." I didn't reveal to her the fish-guts story.

The farm had become rather quiet without the pigs. Feeding the turkeys, chickens, and rabbits had never seemed so simple. It was a little like being an empty-nester: with the kids off to college, thoughts turn toward craft projects and Scrabble. Our house suddenly seemed like a relaxing place to be; it didn't quite feel like a farm anymore, not exactly. But then, I didn't feel crazy anymore, either. And the neighbors seemed awfully relieved.

Chris finished first, and he presented me a perfectly carved coppa heart. We stuffed it into a beef bottom and made coppas. From those trimmings and Big Guy's considerable back fat, we made salami. Chris let me weigh out the spices and use his closely guarded secret recipes for finocchino, soria, and a more basic garlic- and wine-spiked salami.

As we worked, when the pastry chef or a dishwasher came in, he or she invariably praised the slabs of meat that had been Big Guy. I positively glowed. It was better than receiving compliments myself. I was proud of this pig.

Chris's son even came in to admire the pig. Chris pointed out that the meat was pale in color, that it had less myogloblin than most meat. This signaled that it had been raised in a relaxing environment, without being jostled or exercised too much. It's true that my pigs loved their naps, and the only strenuous exercise they got was when I squirted them down with a hose on hot days and they would dance and scamper.

When I went to fetch the machine to stuff my very own salamis, I felt giddy and crazed. Chris had the mixer on, and his son was watching the meat whirl together.

"I want to learn how to make salami," his son confessed.

"You do?" Chris seemed taken aback but pleased.

I loaded the stuffer and cranked out my salamis. Meat, I was glad to see, can really bring a family together.

All these acts—the packing of the salamis, the rubbing of the prosciutto—brought me closer to the pigs. I could see how Big Guy was put together. I knew his secrets.

The most memorable part for me was making the soppressata, or head cheese. After the callousness of Sheila's killing job, the soppressata healed me. To make the dish, Samin put both the pigs' heads in salty water overnight. In the morning, they were drained of blood and looked pale but still wore an expression that I can only describe as optimistic. I imagined that, as they pattered along Sheila's concrete path, they thought they were being taken to an even better place, with even better food.

The heads, with two feet and a tail, plus onions, carrots, and celery, cooked for twelve hours at the restaurant. A full four-inch layer of fat formed at the top of the pot. Chris and I gathered around a table and pulled the meat off the just-warm heads. The fat was still hot.

"Today we are going to re-create that great day!" Chris said. He had already told me the story of a few years ago, when his friend Dario Cecchini, the famous Italian butcher, came to Berkeley for a Slow Food event. Chris had made soppressata. It was a gala affair, with every big-name foodie present and accounted for. Dario, famous for reciting Dante aloud while cooking, attended to the pigs' heads and generally marked the event as epic.

"You know, I'm just disturbed by that woman—what's her name— Sally? Sylvia?—and how she didn't include you," Chris said.

"I know," I said, working on separating out the small bones from a pig's foot.

"It's just wrong," he muttered.

"Why can't people do quality work?" I said.

"Or respect the effort you put into the pigs."

We wiggled open the skulls and scooped out the brains. They were about the size of a large plum. One looked a bit damaged—more like ricotta cheese than brain—so we had to throw it away.

"Taste it," Chris instructed and showed me a good piece of the remaining one to try. I picked up a little and put it in my mouth. It wasn't so much a flavor as a texture. Like a thick piece of cream. Delicious. The unmangled brain went into the bowl with everything else. Chris then added orange peel and an Italian liqueur and stirred up the whole mishmash of fat bits, meat chunks, and carrots. We filled a linen bag with the warm mixture. I was glad that Big Guy and Little Girl would be together again in this dish.

Chris brought down the good Italian meat thread—thick twirls of red and white string. That he was pulling out all the stops on this made me nearly cry. His neat, square fingers snugly tied the bag, then lifted the trays to carry them into the walk-in. There the flavors would meld together for a day or so, the fat would congeal, and thick slices would be cut off and served with cornichons and mustard and crusty bread.

We had used all the parts of the pig, the ultimate show of respect. We spanned time with Big Guy, we pulled off all his flesh, so he could feed us—and feed us very well.

Leslie the pastry chef came in as we cleaned up. She saw Big Guy's picked-over skull on a tray headed for the trash. "Can I have that?" she asked and touched the skull.

"What for?" I asked. I was exhausted from the meat processing.

"I want to mount it on the handlebars of my Schwinn," she said.

These were my people.

However. While I was celebrating the culinary wonders of Big Guy with Chris Lee, other parts of my former pig lingered in the back of my fridge: nameless, uncelebrated, the oozing plastic grocery bag of intestines that Sheila had handed me before I stormed off her property.

After a grueling day of salami stuffing and pancetta rolling, I peered into the fridge and muttered, OK, OK, what is this? From behind a jar of sauerkraut and a half-empty container of stewed plums I pulled out the intestine-filled bag. It visibly quivered. I remembered my indignation, my disgust, at Sheila's waste. Suddenly, it seemed quite reasonable. Ugh, just throw that shit away! But no, between Fergus Henderson and Hugh Fearnley-Whittingstall, I could find recipes for this bag of offal.

Which smelled awful, by the way. I sloshed the bag onto the counter and untied the handles. Inside, cradled by white plastic, lay a liver the size of a placenta. It was a strange red color, almost blue. Next to it sat, rather perkily, a greenish thing that I had to assume was the stomach. I prodded it with my finger, and it yielded only slightly. It had the texture of a scuba diver's wet suit.

The stomach-thing was clearly the source of the overwhelming odor. I smelled my finger: a combo of ripe barnyard and salty low tide. Several hand washings failed to erase the stench from my fingertip. And yet I would transform these things into something delicious.

I gingerly separated the liver from its stinky neighbor and promptly washed it. Then I trimmed all the strange veins and arteries that came and went in and out of the organ. Even after major trimming, I still had two gallons of liver. I cubed it up and, following a recipe called *foie de porc rôti* from Jane Grigson, baked it in a low oven shrouded with a layer of caul fat.

While the liver bubbled in the oven, I turned to the stomach. I like strong-smelling things—ripe French cheeses, deeply fermented cabbages. But every recipe I could unearth for pig stomach involved several days of soaking, often requiring bleach. One that I thought I could get behind, an Amish dish in which the stomach was stuffed with cabbage and sausage and cooked, seemed within the realm of possibility—and edibility. But when I upturned the stomach into our sink for a good scrub before attempting the dish, a green slick that resembled algae flowed from the main orifice. It felt like algae, too, slimy. As the stomach juice slithered down the drain I decided that, in this case, Sheila had been right.

With the stomach back in its bag, I went out to the garden, dug a hole near the base of a struggling fig, and dropped the pig stomach into its final resting place. The bag went into the garbage, where it would undoubtedly become one of the millions of pieces of plastic flying around a landfill somewhere close by. I would not be reusing it.

The liver dish emerged from the oven. I let it cool. It looked beautiful, with the lacing of caul fat dressing up the top of the dish. When I spread some of the dish onto a cracker, it tasted like chalk and blood. It was edible but disappointing: a failure. I somewhat salvaged things by feeding the chickens the *foie de porc rôti*. They went straight to work on it. And though they ate it, I noted that it wasn't with particular gusto.

On the last day of our pig processing, Chris and I went for a walk to scrounge some herbs—wild fennel and rosemary—to stuff into a rolled pork loin. We just walked down to the train tracks, where they were growing like weeds, and clipped fronds here and there.

We passed Chris's Volkswagen van, an unlikely vehicle for the owner of a very fancy restaurant. But then again, Chris was very unlikely. One day when we were talking about lettuces, I told him that I brought salad to the former Black Panthers. Chris got very excited. He grew up in Chicago and was outraged when Fred Hampton—a charismatic young Black Panther leader—was shot in his bed by the police. Although Chris looked as white as the pork back fat, I later discovered that he was part black. His mother was a light-skinned African American who had decided to pass as white. Chris, wanting to get closer to his roots, had looked up the Panthers and what they had been doing and had become politically active himself.

We began picking fennel fronds, and talk turned to urban farming. "I'm not really making a difference," I told Chris. "But Willow and City Slicker Farms, they're doing something that's actually providing people healthy food." Then Chris told me about urban farms that had come before mine. At Chez Panisse, they had relied on urban gardens to grow most of their special lettuces and greens.

"One of the gardens used to be around the corner from a muffler shop," Chris said. "And the lady would arrive—always late—at the restaurant, driving this postal jeep with the greens." Chris paused and smiled at the memory. "We were so cute," he said. "Then we'd wash them very well to get off the muffler smell." There were people raising chickens and bees for honey for the restaurant.

"I think there are townhouses there now," he said when he told me where the urban farm had been.

Back in the kitchen, we washed the fennel well and soaked it in a sink before cutting it finely. Chris found a strange yellow spider on one of the fronds. It was the same color as fennel pollen. While I continued chopping, Chris wandered outside to let the spider go free.

With a giant restaurant skewer, he poked a hole along the bones of the rack of pork loin, and I stuffed it with the freshly cut herbs, packing them in with the handle of a wooden spoon.

How can a restaurant owner be so nice? I wondered as I drove home, the rack of pork loin for twenty in the back of the car. Sure, we had bartered and I had promised to give him one of the prosciuttos made from Big Guy, but this still seemed to be a deal that weighed heavily in my favor. I drove by the intersection where the urban farm had been. Yup, townhouses. They were gray and tall and had lots of parking.

I thought about Chris and his restaurant. From the outside, it was high-end dining, but in the kitchen was a gang of freaks: Leslie the Chinese American pastry chef, who wanted a pig skull on her bike; the pickle-mad Samin; and Chris, onetime radical now teacher. That I was accepted into their tribe made me realize that my identity as urban farmer bridged two worlds, made me an aberration. I might have been a little like the yellow spider Chris saved from the fennel frond.

CHAPTER THIRTY-FOUR

❁

Little Girl's fate had been different from Big Guy's.

I went back up north to pick her up from Jeff's butcher shop, where she had been dismantled in the American style, by a man who had, I couldn't help but notice, a signed letter from President George W. Bush posted on the wall of his meat locker, praising him for this and that. The butcher wasn't there, but his neighbor had come over to let me into the cold room.

"You're from Oakland?" she said. "You've come a long way to get this pig butchered." Like a traitor, I said, "That's because no one in the city knows how to do anything."

She laughed and nodded. After watching Chris trim a pork shoulder of its bone and make soppressata, I knew that city folks know how to do something. I don't know why I said it. In fact, only in the city could I have pulled off feeding my pigs gourmet food. And only in the city were there Italian-trained butchers who were willing to share their knowledge with a novice pig farmer.

The woman wheeled out a dolly stacked high with milk crates filled with pig pieces all wrapped in plastic, then covered with white paper with handwritten labels like BONELESS PORK SHOULDER, GROUND PORK, PORK RIBS. I was thankful for all that meat—it ended up filling an entire upright freezer we borrowed from friends. The butcher had also, per my instructions, saved and wrapped up the bones, the fat pieces, the feet, and the trimmings.

All that wrapping, though, didn't have much soul. The butcher had used a band saw to take the pig apart, so the meat left few hints about pigness— the lines were straight, not organic; square-shaped. Little Girl ceased to exist.

But Big Guy, in the form of his giant hanging butt, had become immortal, in a way.

This soullessness is not how it has always been in America. On Samin and Chris's advice I had read Edna Lewis's essay "Morning-After-Hog-Butchering Breakfast," an elegy about traditional Southern hog butchering. After the slaughter, cleaning, and hanging of the pigs in the December cold, Lewis remembered, "we waited with impatient excitement through the three days of hanging; we were all looking forward to the many delicious dishes that would be made after the hogs were cut up—fresh sausage, liver pudding, and the sweet delicate taste of fresh pork and bacon."

I had also recently run across a copy of *Little House in the Big Woods,* by Laura Ingalls Wilder, which fueled this nostalgia. I decided I would try some typically American moves with Little Girl. So I took one of the hams, unwrapped it from its swaddling of paper and plastic, and submerged it in a brining solution for ten days. I released a slab of belly and began the process of making bacon: I cut the big slab into three smaller sections and rubbed them with a little pink salt, kosher salt, pepper, and maple syrup, then placed them in the fridge to marinate for a few weeks, flipping them every other day to spread the brine evenly.

And so Big Guy and Little Girl were still alive, in my heart and in my daily to-do list, just like at a Southern hog butchering, where Edna Lewis remembered that "as soon as the hogs were butchered, a series of necessary activities ensued that kept the whole community busy for at least a week or more." I made almost daily trips to the restaurant to flip prosciuttos, roll and hang pancettas, and monitor the salamis made from Big Guy. I also checked on the Little Girl ham and checked on the belly bacon every day.

I did often wonder: Why did Chris Lee help me?

We talked about it while making lardo, our last task.

"You know, you're really lucky," said Samin.

We were watching Chris remove some skin from a piece of back fat that we were going to rub with salt and fennel and hang in the walk-in.

"I know," I said. "Look at me, I'm an asshole, just watching Chris do all the work."

"You said the magic words," Chris said.

"What did I say?" I asked.

"You said you had two pigs."

"Chris hates it when people come to the front of the restaurant and want to sell him stuff," Samin explained.

"But you said you had two pigs," Chris said, "so I had to talk to you."

Chris told me that he had thought I was some kind of rich-lady hobby farmer who lived in some rural area and wanted advice on raising pigs. When he met me and realized I was some poor hobby farmer from the ghetto, he was intrigued.

I had to give something back. One day I slipped the pastry chef a wrapped package of Little Girl's feet. She yelled in delight: She was going to make a special stuffed Chinese pork recipe that her grandmother used to make.

I gave another chef some pork bones to take home to braise and make into a stock—something I had done at home with delicious results. Never underestimate the deliciousness of pork stock. Since it was getting to be fall, the stock was genius for making soups and stews. The baking bones also filled our house with wonderful odors. I tried to give Chris's son good advice on cool colleges for wee punk rockers—like Reed in Portland or Evergreen in Olympia. I promised Samin I would help her with her writing. As for Chris, I figure I still owe him one. A big one.

In the middle of the pig mania, Dante called me on my cell phone.

"Hi. I wanna buy a rabbit," he said.

I was riding my bike to the restaurant and nearly fell off.

"Who's this?" I asked, thinking Bill was playing a joke.

"The kid who came over with my friends and we saw the pigs," he said. Then I heard the phone clattering and a woman's voice came on. "I'm Dante's mom, Gwen," she said.

"Hi, Gwen. Is it OK for Dante to get a rabbit?" I asked.

"Yes. He's been talking about it for months," she said. "And he saved up his money." Dante, it turned out, was a little entrepreneur who funded his own cell phone, clothes, and hairdos.

We made arrangements to meet.

The whole family showed up—Dante and his mom, sister, and brother.

I took them out to the deck so he could pick his favorite rabbit. After cradling all the contenders, he chose a soft, light brown male. I gave him an extra cage I had and a small bag of food. His brother and sister also held rabbits.

"I'm going to save up and buy one, too," Dante's brother said.

I ended up charging Dante $5. I told him how to hold the rabbit, how to change the bedding, how to fill the waterer. He nodded when I told him all these things. He was a sharp kid.

I knew that Dante just wanted a pet, but I couldn't help but see him as a future urban farmer. As he and his family walked home I watched them from the deck with a sense of pride I couldn't explain to myself.

As the days grew short and the nights cold I fired up my barbecue and sprinkled it with wet hickory chips. When the white smoke billowed up, I placed my slabs o' bacon on the grill and left them to inhale smoke for hours.

Then, after another day in the fridge to firm up, I cut a few slices and fried them for breakfast. Homemade bacon is nothing like store-bought. Mostly because it just isn't perfectly square. Bacon factories, it turns out, use molding machines to convince the meat to form a perfect rectangle. My knife wasn't very sharp, either, so the slices were fairly large.

As I sat down to breakfast that morning with a gleam in my eye, poised to put a wavy piece of pork belly into my mouth, I thought of my Las Vegas bacon-eating frenzy oh so long ago. How could I have known that I would end up here, exhausted by five months of pig-raising effort and finally getting to eat something that I had transformed from mundane to extraordinary? I sank my teeth into the unctuous fat, the crispy meat. It was just as bacon should be—smoky and sweet, salty and peppery.

Bill walked into the kitchen, sleepy, his hair tousled. He had been bummed that we had to wait a few months before we could cut into the salamis and slice up the pancettas that were hanging in Chris's walk-in. But the bacon was ready to go.

"How is it?" he asked.

I shrugged. "Pretty eff-ing good."

He chomped on a piece, then wolfed it down and grabbed another. We were finally enjoying the fruits of our labors. The hard work of feeding two pigs had paid off. Bill and I kissed in celebration, both of our mouths salty and sweet from the bacon. I fired up the cast-iron pan again to make more bacon, and as it fried, filling our kitchen with the scent of sizzling pork, we never felt so lucky. We had achieved the heights of urban farming together. The meat—it was official—was amazing. God, we were unbearable.

CHAPTER THIRTY-FIVE

✵

In October, well after all the pork was salted away and I had stopped visiting the restaurant, a sign went up in front of the squat garden. FOR SALE, it read. There was a phone number, and I called it. They would like to sell it for $488,000, the agent told me. I laughed at this price and told her it was a lot in the middle of the ghetto. I didn't tell her who I was.

"Well, that area is in transition," she said.

It was true. I had just the other day spotted a team of ultimate-Frisbee players in the abandoned schoolyard. A group of artists had moved into Lana's warehouse and turned it into an art gallery. And a Whole Foods had just gone up a few blocks away.

"It's zoned commercial," she told me. "You can put condos up." Assuring her that I didn't have that kind of money, and if I did, I wouldn't build condos, I hung up, shaken.

By the winter solstice, the sign still hung in front of the lot. Someone—not me—had knocked it halfway down. To celebrate another year almost over, I threw a party. I served the salami and lardo that I had made with Chris, and the wine I had left from the winemaking party with Jennifer and Willow.

There's a Portuguese saying: The happiest times in life are the first year of marriage and the week after you butcher a pig. The bounty had been overwhelming, and the happiness extended to months. There were sixteen salamis, four pancettas, two coppas, four lardos, and two prosciuttos hanging in the walk-in at the restaurant.

As they were ready I would take them home to eat—and distribute. My prosciutto, Chris agreed, could stay at the restaurant until it was ready. In our own freezer were the American cuts: pork chops, ground pork, spareribs, and pork back fat. We had another side of pork belly to make bacon with, and another ham—the first one had been outrageous. We had hosted six dinner parties over the past few months, one featuring a banana-leaf-wrapped pork loin, another with pork tacos from slow-roasted spareribs. We had even hosted a sausage-making party.

This was the party where we would debut the salami and the lardo, which had taken three months to cure in Chris's meat room. He had devoted a special section of the meat cave to my meats. I even made a special tag for it: a purple N highlighted by yellow marker.

I had sent my mom, for the solstice, a fennel-spiked salami. She sent me an e-mail raving about it. "I'm slowly slicing off pieces, savoring it," she wrote. I gave Mr. Nguyen a slab of pork ribs. I sent my sister the most quintessential American product: leaf lard. I had rendered pieces of back fat over one day, slowly draining off the fat, which melted on a low flame. It was pure white, like porcelain. I was thankful for sharing, for redistributing the pork. Otherwise, I was going to balloon up just like one of the pigs.

For the party plate, I sliced everything as thin as possible. Whispers of salami, slivers of lardo. The guests packed into our cramped kitchen and snarfed up the food. I couldn't decide what made me happier: having seen the pigs eat so happily, months ago, or watching my friends do so now. My salami, I thought, was as good as Chris's—the fennel seeds shined through and blended with the meat flavor perfectly. Little cracks of pepper and hot paprika in the other dazzled our palates. The lardo—cool, salty, sweet—soothed the heat.

A few guests wanted to hear how I killed rabbits on my farm, and so I narrated their death. I also told the story of how I met Chris for the hundredth time and, like my mother, never grew tired of the telling. We poured the homemade wine and offered guests tastes of the fall honey harvest. I felt slightly embarrassed at the riches in our larder.

By midnight, almost everyone had gone home. A few late-nighters drank the last of the wine and considered our couch.

I walked outside to feed the rabbits, my usual ritual before turning in. The deck quivered with activity. I tossed some bok choy salvaged from the Dumpster from a bucket. The rabbits pounced and nibbled on the greens. Going to the Dumpster now, postpig, was a rather sad exercise. We left so much more than we could take.

I was a little drunk and felt a bit melancholy. At one point the deck had been a hangout for humans. Then it hosted bees and a container garden. At the moment, it was a rabbitry. I loved this place because from here I could view our whole street—the hustlers and the artists, the families with their struggles.

A neighbor turned the corner holding a black bag filled with beer. Joe and Peggy were taking their dog out for a walk. The monks had been preparing a feast all day. The garden was pensive this winter. Someone had set up a table covered with stuffed animals and baby clothing marked with a FREE sign just off MLK, right next to the battered FOR SALE sign.

These past few years had been strange ones, perhaps, for a place known as GhostTown. All of us—the Vietnamese families, the African American teenagers, the Yemeni storekeepers, the Latino soccer players, and, yes, the urban farmers—had somehow found a way to live together. To share and discover our heritage with one another. But now I could feel that an end, or a change, was afoot in this almost new year.

People will come and go. Animals will be birthed and die. Food and flowers will be plucked from the earth, friendships made. Bullets will be fired. Houses will be boarded up, then sold to be fixed up. Innumerable sodas and malt liquors will be purchased from Brother's Market, and many of them will be consumed in the street. Weeds will feed animals that will then feed humans. Dice will be thrown. Children will grow up and move few blocks away from their parents' house. Incense will be burned, fireks set off, trash hurled from a moving car. A man will start a new life in he can call home. A grandmother will sell dinners of fish she caught oked herself. Looking back on it, we in this neighborhood were all ns of a sort. No one would have bet on any of us.

playing the part of an undertaker again. The body before me was that of an urban farm. Before long, I imagined, I would leave it, with more nutrients, more plants in the soil for the bulldozers to unearth. But in leaving it, I would take it with me, too. Not just in my body, which had ingested its riches and grown strong in the working of the farm, but in my spirit—all the things I had learned, my singing heart, my smile lines, my aching bones. I hadn't truly owned any of this place. It had owned me.

And now I was just one of the many ghosts in GhostTown. I sprang up here only because it was the perfect intersection of time and place, and, like a seedling, I took advantage, sucked up the nutrients that I could find, forged relationships with others in order to grow, bathed in the sunlight of the moment.

I had been lucky during these past years. Somehow, all the forces had aligned to make my life full and abundant. I had arrived at a time when an abandoned lot could be taken over, a backyard turned into a place to keep animals, connections between humans made. This time had now passed.

My farm will eventually be bulldozed and condos will be built. Bill and I will move somewhere else. Where, undoubtedly, we will first build the garden. Then set up a beehive. Then chickens . . . Being part of nature connected us to the past, the present, and the future.

And who knows, maybe a few neighborhood kids like Dante will pass by the units and tell someone who doesn't care, "There used to be a farm here." Maybe the peach trees planted in the parking strip will remain, and a hungry urban forager will cherish the ripe peaches someday. The soil here will be uncommonly abundant, and maybe someday a strange-looking vegetable will sprout here again, when the moment is ripe.

CHAPTER THIRTY-SIX

One of Willow's chickens turned out to be a rooster. We had distributed most of the chickens to her backyard-garden people, but I still had a few, including the rooster. He was beautiful, with red glossy chest feathers and giant bobbing green tail feathers. At first I enjoyed the crowing, but then I noticed it was happening around 3:30, 4 a.m. It was never just one crow, either—it was over and over again.

In any case, it was extremely annoying. And dangerous. If my neighbors complained, who knows, maybe animal control would come and take my chickens and rabbits. Thinking about my options, I rode my bike past Brother's Market.

"Hey-hey," Mosed the shopkeeper yelled.

I slowed down and peeked in.

"Where's my honey?" he asked.

"I've got some for you—I've just been busy," I answered. We had harvested a boatload in the fall. "Want a rooster?"

He came outside. His dyed red hair sparkled in the weak December sun. He nodded. Tomorrow, I told him, I'd bring him the rooster and some honey.

The rooster slept outside the henhouse, protecting his ladies, I guess. I nabbed him in the morning, tossed him in a cage, and walked half a block to Mosed's market. The rooster had already put in a few crows before 8 a.m.

Inside the store, filled with malt liquor and chips, a woman wearing a head scarf sat on a chair peeling an orange. When she saw me, she let out a torrent of words. The customer in line did a double take at the rooster, then gathered his black plastic bag of beer and left.

I set the cage on the ground. Mosed came around to look at the rooster. I handed him the jar of honey. He smiled. "How much?" he asked.

"Ten dollars for the rooster. The honey's a gift."

Mosed went back around to the cash register and opened the till. His wife shouted a few words, ate a slice of orange.

"She thinks that's too much, huh?" I said. A woman's displeasure is apparent in any language.

"Yes, but don't worry about it," he said. To make her feel better, Mosed showed her the honey. He waved the jar in front of her until she took it out of his hands.

I looked down at the rooster. I was sure Mosed would do a better job dispatching this guy than I would.

Then I was walking home in the cold December air, the sun suddenly bright, a well-worn GhostTown ten-dollar bill in my pants pocket. I wanted to tell Mosed that I had finally figured out who I was, who my people were: they were folks who love and respect animals, who learn from them, draw sustenance from them directly.

Although my holding was small—and temporary—I had come to realize that urban farming wasn't about one farm, just as a beehive isn't about an individual bee. I thought of Jennifer's beehive and garden. Of Willow's backyard farms that dot the city of Oakland. Urban farms have to be added together in order to make a farm. So when I say that I'm an urban "farmer," I'm depending on other urban farmers, too. It's only with them that our backyards and squatted gardens add up to something significant. And if one of ours goes down, another will spring up.

Now facing eviction and change, which is always part of our shifting city life, this time it was my farm that would go under. It was sad, yes, but I knew that wherever I went I would continue to grow my own food, raise animals, love and nurture life in places people thought were dead. And if anyone asked, I could say: I am a farmer.

ACKNOWLEDGMENTS

For helping me through the awkward molting stages of this project: Russ Rymer, Lisa Margonelli, Heather Smith, Lygia Navarro, Morgen VanVorst, Peter Alsop, Zach Slobig, Chris Colin, Kate Golden, Kristin Reynolds, Sue King, and Traci Vogel, I thank you.

I would not be an urban farmer without the help of my fellow agriculturalists Willow Rosenthal, Severine von Tscharner Fleming, Jim Montgomery, John White, and Jennifer Radtke. For showing me how to respect pork, huge hugs to Chris Lee and Samin Nosrat. Thanks to my ever-patient neighbors, especially Lana, the monks at the Kwan Yin Temple, BBob, Logo, and the Nguyen family.

Much love to my father and mother; to my sister, Riana; to Benji Lagarde and his family; and to the future: Amaya Madeleine Lagarde. Lawry Gold, I named my hard drive after you.

I'd like to thank my copy editor, Candice Gianetti, and production editor, Bruce Giffords, for their careful work. At The Penguin Press, my editor, Jane Fleming, and publisher, Ann Godoff, were a dream team—thanks for believing in me. Michelle Brower, my agent, has been a source of delight.

Billy Jacobs, you are the best reader—and boyfriend—a gal could have.

And finally, thanks to the farm animals: *Mellagris gallapavo, Gallus domesticus, Sus scrufo, Anas domesticus, Oryctolagus cuniculus,* and *Apis mellifera.* I couldn't have done it without you.

BIBLIOGRAPHY

BARNYARD

Bairacli Levy, Juliette. *The Complete Herbal for Farm and Stable.* London: Faber & Faber, 1952.

Bennett, Bob. *Raising Rabbits Successfully.* Charlotte, VT: Williamson Publishing, 1984.

Cato, Marcus Porcius. *Roman Farm Management: The Treatises of Cato and Varro.* Middlesex, UK: Echo Library, 2007.

Dadant, C. P. *First Lessons in Beekeeping.* Hamilton, IL: Great River Publishing, 1917, 1924, 1976.

Damerow, Gail. *Barnyard in Your Backyard: A Beginner's Guide to Raising Chickens, Ducks, Geese, Rabbits, Goats, Sheep and Cattle.* North Adams, MA: Storey Publishing, 2002.

Emery, Carla. *The Encyclopedia of Country Living: An Old-fashioned Recipe Book.* Seattle: Sasquatch Books, 1994.

Evans, Robert Jones. *History of the Duroc.* Chicago: James J. Doty, c. 1918.

Holderread, Dave. *Storey's Guide to Raising Ducks.* North Adams, MA: Storey Publishing, 1978.

Horn, Tammy. *Bees in America: How the Honey Bee Shaped a Nation.* Lexington: University of Kentucky Press, 2005.

Hubbell, Sue. *A Book of Bees . . . and How to Keep Them.* New York: Ballantine Books, 1988.

Maeterlinck, Maurice. *The Life of the Bee.* New York: Dodd, Mead, and Company, 1913.

Olkowski, Helga. *The Integral Urban House: Self-Reliant Living in the City.* San Francisco: Sierra Club Books, 1979.

Root, Amos Ives. *The ABC and XYZ of Bee Culture,* 40th ed. Medina, OH: The A. I. Root Company, 1990.

Smith, Andrew. *The Turkey: An American Story.* Chicago: University of Illinois Press, 2006.

Towne, Charles Wayland, and Edward Norris Wentworth. *Pigs: From Cave to Corn Belt.* University of Oklahoma Press, 1950.

Van Loon, Dirk. *Small-Scale Pig Raising.* Pownal, VT: Garden Way Publishing, 1978.

GARDEN

Ashworth, Susan. *Seed to Seed: Seed Saving and Growing Techniques for Vegetable Gardeners.* Decatur, IA: Seed Savers Exchange, 1992.

Biggs, Matthew. *Matthew Biggs's Complete Book of Vegetables: The Practical Sourcebook to Growing, Harvesting and Cooking Vegetables.* London: Kyle Cathie, 1997.

Flores, Heather C. *Food Not Lawns: How to Turn Your Yard into a Garden and Your Neighborhood into a Community.* White River Junction, VT: Chelsea Green Publishing, 2006.

Goldman, Amy. *The Compleat Squash: A Passionate Grower's Guide to Pumpkins, Squashes, and Gourds.* New York: Workman Publishing, 2004.

Guillet, Dominique. *The Seeds of Kokopelli.* Boston: Kokopelli Seed Foundation, 1981.

Hartman, Hudson. *Plant Propagation: Principles and Practices.* Upper Saddle River, NJ: Prentice Hall, 1997.

Hemenway, Toby. *Gaia's Garden: A Guide to Homescale Permaculture.* White River Junction, VT: Chelsea Green Publishing, 2003.

King, F. H. *Farmers of Forty Centuries: Organic Farming in China, Korea, and Japan.* Mineola, NY: Dover Publications, 2004 (first published in 1911).

Pack, Charles Lathrop. *War Garden Victorious.* Washington, D.C.: National War Garden Commission, 1919.

Spirn, Anne Whiston. *The Granite Garden: Urban Nature and Human Design.* New York: Basic Books, 1984.

Tucker, David. *Kitchen Gardening in America.* Ames: Iowa State University Press, 1993.

Warner, Sam. *To Dwell Is to Garden: A History of Boston's Community Gardens.* Boston: Northeastern University Press, 1987.

Wright, Richardson. *The Story of Gardening: From the Hanging Gardens of Babylon to New York.* New York: Dodd, Mead, and Company, 1934.

KITCHEN

David, Elizabeth. *French Provincial Cooking.* New York: Harper & Row, 1960.

Fearnley-Whittingstall, Hugh. *The River Cottage Meat Book.* Berkeley: Ten Speed Press, 2007.

Fisher, M.F.K. *How to Cook a Wolf.* San Francisco: North Point Press, 1988.

Grigson, Jane. *Charcuterie and French Pork Cookery.* London: Penguin Books, 1970.

Henderson, Fergus. *The Whole Beast: Nose to Tail Eating.* New York: HarperCollins, 2004.

McGee, Harold. *On Food and Cooking: The Science and Lore of the Kitchen.* New York: Scribner, 1984.

O'Neill, Molly, ed. *American Food Writing: An Anthology.* New York: Literary Classics of the United States, 2007.

Ruhlman, Michael, and Brian Polcyn. *Charcuterie: The Craft of Salting, Smoking and Curing*. New York: W. W. Norton, 2005.

Visser, Margaret. *The Rituals of Dinner: The Origins, Evolution, Eccentricities and Meaning of Table Manners*. New York: HarperCollins, 1991.

LIBRARY

Agnew, Eleanor. *Back from the Land: How Young Americans Went to Nature in the 1970s, and Why They Came Back*. Chicago: Ivan R. Dee, 2004.

Allport, Susan. *The Primal Feast: Food, Sex, Foraging, and Love*. New York: Harmony Books, 2000.

Berger, John. *About Looking*. New York: Vintage, 1991 (first published 1980 by Pantheon Books).

Berry, Wendell. *The Unsettling of America*. San Francisco: Sierra Club Books, 1977.

———. *What Are People For?* New York: North Point Press (FSG), 1990.

———. *The Art of the Commonplace: The Agrarian Essays of Wendell Berry*. New York: Shoemaker & Hoard, 2003.

Brand, Stewart. *Whole Earth Catalog*. Berkeley: Whole Earth Publishing, 1968.

Budiansky, Stephen. *The Covenant of the Wild: Why Animals Chose Domestication*. New York: William Morrow & Company, 1992.

Cronon, William. *Nature's Metropolis: Chicago and the Great West*. New York: W. W. Norton, 1991.

Fernald, Anya; Serena Milano; and Piero Sardo, eds. *A World of Presidia: Food, Culture, and Community*. Bra, Italy: Slow Food Editore, 2004.

Gibbons, Euell. *Stalking the Wild Asparagus: Field Guide Edition*. New York: David Mackey, 1970.

Hough, Michael. *City Form and Natural Process: Towards a New Urban Vernacular*. London: Routledge, 1984.

Johnson, Marilynn. *The Second Gold Rush: Oakland and the East Bay in World War II*. Berkeley: University of California Press, 1993.

Lawson, Laura. *City Bountiful: A Century of Community Gardening in America*. Berkeley: University of California Press, 2005.

MacDonald, Betty. *The Egg and I*. New York: Harper & Row, 1945.

Pellegrini, Angelo. *The Unprejudiced Palate: Classic Thoughts on Food and the Good Life*. San Francisco: North Point Press, 1984.

Pinderhughes, Raquel. *Alternative Urban Futures: Planning for Sustainable Development in Cities Throughout the World*. Oxford: Rowman & Littlefield, 2004.

Plath, Sylvia. *Ariel*. New York: HarperCollins, 1963.

Rhodes, Jane. *Framing the Black Panthers: The Spectacular Rise of a Black Power Icon.* New York: The New Press, 2007.

Rorabaugh, W. J. *The Alcoholic Republic: An American Tradition.* London: Oxford University Press, 1979.

Shepard, Paul. *Thinking Animals: Animals and the Development of Human Intelligence.* Athens: University of Georgia Press, 1978.

Thoreau, Henry David. *Walden.* New York: W. W. Norton, 1992.

Vileisis, Ann. *Kitchen Literacy: How We Lost Knowledge of Where Food Comes From and Why We Need to Get It Back.* Washington, D.C.: Island Press, 2008.

Wilder, Laura Ingalls. *Little House in the Big Woods.* New York: HarperCollins, 1935.